"In a series of engaging stories, Brown tells about his life as a cook. . . . Each story has the feeling of a journey from ignorance (or just plain insensitivity) to enlightenment, and each is followed by a little bouquet of recipes that, miraculously, reflects this enlightenment."

—*San Francisco Chronicle*

"If you have any doubt about choosing between a self-help book and this one, don't hesitate to learn your radish teachings."

—Thomas Moore,
author of *Care of the Soul* and
The Re-Enchantment of Everyday Life

"An inspired collection of delicious, doable recipes, ones that I look forward to cooking in my own kitchen."

—Patty Unterman,
owner of Hayes Street Grill and
author of the *San Francisco Examiner*'s
"Dining Out" column

Praise for The Tassajara Bread Book:

"This was the first cookbook I ever bought for myself. . . . To this day, I consider [it] to have been a major influence not just on my cooking and baking, but on my attitude and philosophy about food in general. Thank you, Ed Brown, for this lasting gift."

—Mollie Katzen, author of
The Moosewood Cookbook and
The Enchanted Broccoli Forest

"The bible for breadmaking."

—*The Washington Post*

Praise for Tassajara Cooking:

"Rarely has a book of such simplicity underscored so well the joy of culinary discovery."

—*Bon Appétit*

Also by Edward Espe Brown

The Tassajara Bread Book

Tassajara Cooking

The Tassajara Recipe Book

The Greens Cookbook (with Deborah Madison)

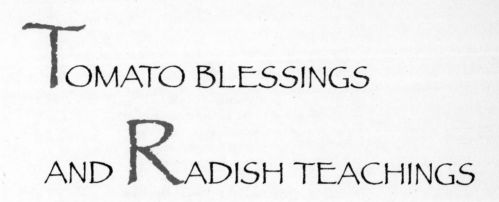

TOMATO BLESSINGS
AND RADISH TEACHINGS

Recipes and Reflections

Edward Espe Brown

RIVERHEAD BOOKS, NEW YORK

RIVERHEAD BOOKS
Published by The Berkley Publishing Group
A member of Penguin Putnam Inc.
200 Madison Avenue
New York, New York 10016

First Riverhead hardcover edition: April 1997
First Riverhead trade paperback edition: July 1998
Riverhead trade paperback ISBN: 1-57322-673-4

The Penguin Putnam Inc. World Wide Web site address is http://www.penguinputnam.com

The Library of Congress has catalogued the Riverhead hardcover edition as follows:

Brown, Edward Espe.
Tomato blessings and radish teachings :
recipes and reflections / Edward Espe Brown
p.   cm.
ISBN 1-57322-038-8
1. Cookery. 2. Cookery—Religious aspects—Zen Buddhism.
I. Title
TX652.B72   1997
641.5—dc20   96-43994 CIP

Printed in the United States of America

10 9 8 7 6 5 4 3 2 1

This book is dedicated to fathers
and to fathering, which of course
includes a lot of mothering.

Especially it is dedicated to the
memory of my blood father,
Frank Espe Brown,
and my spiritual ancestor
Shunryu Suzuki.

Acknowledgments

Each of us is who we are largely because of others. My life manifests itself as it does because others have believed in me, which has allowed me to believe in myself. This has been true since early in my life, starting with my parents, who also gave me a love of language and ideas. As for my capacity to write, I am fortunate to have had a number of most excellent teachers, especially in English departments, all through high school and for a year at San Francisco State. The results, along with my gratitude, speak for themselves.

I have been working on this book, *Tomato Blessings and Radish Teachings*, for more than ten years. Versions of some of these pieces and recipes appeared first in the *Yoga Journal*, thanks to Stephen Bodian, who was editor at the time, and more recently in the *Shambhala Sun*, thanks to Melvin McLeod, the *Sun's* editor. Their appreciation of my writing was a strong encouragement to continue, so I am indebted to the two of them. These magazines, by the way, are both interesting and instructive.

Sam Bercholtz of Shambhala Publications looked at my manuscript copy of *The Tassajara Bread Book* for a few minutes in the basement of the Shambhala Bookstore in 1969 and agreed to publish it. I am grateful to Sam for his vision of what that book could be, and I am certain that his sense of design, layout, illustrations, and calligraphy had a great deal to do with the book's success. Later Shambhala also published *Tassajara Cooking* and *The Tassajara Recipe Book*. Some recipes from those books have been adapted for this one. Shambhala has continued to publish a wide range of absorbing books.

Innumerable cooks, many of whom are mentioned in the text of this book, have been an encouragement to me over the years. Here I would like to express my appreciation of Alice Waters and Deborah Madison in particular. Alice's down-to-earth love for the ingredients themselves and her encouragement of growers and producers have been simply phenomenal, and I have always found her to be one of the most generous people I know. Deborah has a highly developed, poetic sensibility about food and cooking, and assisting her on *The Greens Cookbook* was an empowering experience for me. Some of the recipes in this book are variations of recipes which appear there. *The Greens Cookbook* and her other book, *The Savory Way*, have encouraged many people in their development as cooks. My thanks and gratitude to Alice and Deborah, and best wishes for their happiness and well-being.

My friend and agent, Michael Katz, has had an invaluable and irreplaceable part throughout the whole process of this book's creation. It simply would not have happened without his input, which included leaving me alone for long periods of time with a terse instruction to "Keep working," or a suggestion like, "You need a Table of Contents." Eventually things unfolded. I find visiting with Michael to be provocative and energizing, and I am terribly relieved and most wonderfully blessed that he survived an emergency surgery to repair a tear in his aorta in February. A warm place in my heart for you, Michael.

Michael also forged my connection with Riverhead Books. My thanks to Susan Petersen for her enthusiastic appreciation of my writing, and to Mary South, my editor, for respecting my voice as a writer. Also, Mary's assistant, Kathryn Crosby, was a helpful and friendly presence throughout. May your endeavors flourish.

Working closely with me on the manuscript was Jisho Warner, who has a fine sensibility, good humor, and a fondness for food. She got me to trim adjectives from five or seven to three or four, and suggested numerous other beneficial prunings. Much of my writing gets tangential and redundant, and if the book now seems cohesive, we have Jisho to thank. Intrepidly and delicately she entered the world of my writing, and we both survived. My heartfelt gratitude.

Acknowledgments

I am also especially pleased and thankful to David Bullen for his cover and design work.

A number of people have been particularly instrumental in encouraging me to write, including my mother, Anne, and my friend Nicole Young. This has been essential because I do not really identify with being a "writer" and would not be writing if no one expressed an interest. I also appreciate the support and encouragement from George Lane, Ginny Stanford, Gary Thorp, and Ann and Gene Swett of the Old Monterey Inn. Thank you one and all.

Abbot Sojun Mel Weitsman of the Zen Center of San Francisco spent many hours going over the manuscript with me, making several beneficial suggestions including adjusting Suzuki Roshi's language here and there. Abbot Zoketsu Norman Fischer read parts of the manuscript with evident joy and relish which buoyed my spirits.

Finally I am grateful to my companion, Patti Sullivan, for her love, and also for her cooking and the occasional recipes that she contributed to this book. Patti is more interested in her artwork than in cooking, but over our years together she has become quite a capable cook, and it's nice not having to cook all the time.

May all beings benefit.

Edward Brown
Inverness, California
October, 1996

Contents

Perfect apple, pear, and banana,
gooseberry . . . All of these speak
death and life into the mouth . . . I sense . . .
Read it in the face of a child

who is tasting them. This comes from far away.
Are names slowly disappearing in your mouth?
In place of words, discoveries are flowing out
of the flesh of the fruit, astonished to be free.

Dare to say what it is we call "apple."
This sweetness, compressed at first,
then, gently unfolded in your tasting

becomes clear, awake, and transparent,
double-meaninged, sunny, earthy, here –:
Oh, realizing, touching, joy –, immense!

> from *The Sonnets to Orpheus* by Rainer Maria Rilke
> version by Hermann Clasen and Ed Brown (unpublished)

Introduction: Cooking and Spiritual Practice

I began cooking about the same time I began practicing zen meditation in 1965. These two endeavors have so enriched my life over the years that by now they seem inseparable. I love the interplay between formal spiritual practice and everyday life. After all, this is where we live: with things that are not just things, and with meaning that can be more real than things. I want the spiritual to reach the kitchen. Otherwise it is empty of significance. What's for lunch?

Cooking is not merely a time-consuming means to an end, but is itself healing, meditation, and nourishment. At one point my Zen teacher Shunryu Suzuki Roshi commented, "You are not just working on food, you are working on yourself, you are working on other people." We sell ourselves short when we concentrate on instant relief and instant gratification and do not see that work is how we make our love manifest.

By cooking we share in the effort that sustains our lives; we share in the bountifulness of our world. Onions and potatoes, apples and lettuces are intimately interconnected with earth and sky, sun and water. Giving and receiving, they do their best.

So handling food can also be an expression of sincerity and appreciation, working with so many living things. Honoring the ingredients, we can work to bring out the best in the food, in ourselves, and in one another.

Many of our individual and social difficulties come from arbitrarily compartmentalizing our lives. Work is what we do for a living, and it is so often unrewarding that it becomes something to be avoided. Cooking provides a metaphor and a vehicle for making our lives whole: finding pleasure in handling food instead of thinking that pleasure comes when the work is over. Experiencing the simple joys of creating with wheat and corn, tomatoes and arugula, instead of thinking that joy is not having to relate to anything.

We have forgotten what really nourishes us, and when we fail to connect with things, life becomes empty and deadening. To see food merely as fuel or stuff is impoverishing. Enlightenment or realization in Zen is sometimes referred to as "attaining intimacy": It is to actually touch and know through and through, to digest and grow. We cannot be more intimate than we are with food; it becomes us. The so-called spiritual cannot be separated from the so-called material. Spiritual food is right at hand, but we have to pick it up, smell it, and taste it.

Cooking, then, is life, is learning, is realization, but, hey, let's not just talk about it, let's eat. And let's not be mindless about what we're putting in our mouths. Let's taste the incredible blessing in each bite of food, which is also the blessing of our ability to taste and be aware, the blessing of our capacity to grow and nourish one another.

ON THE IMPORTANCE OF HAVING FIASCOES

I am firmly convinced that having fiascoes is an unavoidable part of learning to cook, a pivotal aspect of growing up. If you want to be a cook, if you want to come into your own, you will meet fiascoes along the way.

Rain falls at an outdoor wedding, a batch of cookies burns, dinner is late, relationships sour or turn as life unfolds. Common myth has it that if we behave correctly, follow the directions, behave by the rules, everything will be OK. Cookbooks often say this – that if you do what we say, everything will come out great. Yet circumstances arise that are beyond anything we could have imagined.

The fiasco may feel like a betrayal: "How could you do this to me?" we ask the universe. "Why me?" Or we find fault and assess blame: Someone did not follow the rules; they didn't do what they should or could have done. Sometimes we decide to never try particularly hard again, so we will never have to encounter another inexplicable failure after making a tremendous effort.

Yet our capacity to endure fiascoes is what allows us to realize our own treasures. Our straightforward successes do not call forth our deep resourcefulness. Coming up short can bring out our vitality and ingenuity, our generosity and patience. Perhaps everything will not come out perfectly, but we'll do what we can, fully.

After several years of study, a student leaving Tassajara offered a few words of advice: "Don't be afraid of difficulty," he said. "The Dalai Lama says that you learn the most from difficulty." He paused before concluding, "I think you can trust those words."

Just so we understand one another, I don't mean endlessly staying in an abusive or unworkable situation, thinking, "This must be good for me." To stand up and walk away may be what is learned. Difficulty can help you come into your own. And once you act, help will find you.

A PATH IN LIFE

This is a book about my path through life. Along the way I thought I was going here or going there, doing this or that, yet only now can I see how many of the threads connect. There's a profound mystery to this raveling and unraveling, to the unfolding of each life.

After discovering zen practice and cooking in 1965, I got a job as the dishwasher at Tassajara Hot Springs in May 1966. A summer resort since the 1880s, Tassajara was purchased by the San Francisco Zen Center in December of that year. In its last year under the ownership of Bob and Anna Beck, I apprenticed in the kitchen, not only washing the dishes, but learning to make bread, assemble soups, and prepare breakfast. Jimmie Vaughn and Ray Hurslander were the cooks who tutored me. I owe a great deal to their instruction and encouragement.

Halfway through that summer a pivotal event occurred that gave my life a direction. Ray quit, and I was offered his job. Accepting this position I encountered the twin challenges of learning to cook and learning to cope with cook's temperament. Jimmie continued as the dinner cook and gave me guidance as I undertook the task of preparing breakfast and lunch. Lunch was especially useful training, since it involved using up leftovers.

The following May I accepted an invitation to be the head cook of the new Zen Mountain Center at Tassajara, and I continued in this position through the summer of 1969. Many of the stories in this book are from that pivotal period in my life.

My first book, *The Tassajara Bread Book*, came out in 1970. Published by Shambhala Publications, the *Bread Book* was christened the "bible of bread-making," by the *Washington Post* and became an instant best-seller, contributing, it seems, to the revival of bread-making in this country.

In 1971 I was ordained as a Zen priest by Shunryu Suzuki Roshi and given the Buddhist name Jusan Kainei, Longevity Mountain Peaceful Sea. Meanwhile in the summer of 1970 I married Meg Gawler, and our daughter Lichen was born in April 1973, the year my second book, *Tassajara Cooking*, appeared. Having arrived at Tassajara with almost nothing in 1967, by the time I left in December 1973, I had a wife, a daughter, a car, a carload of possessions, and two books to my name. Zen in America is not like Zen anywhere else in the world.

We moved to the Zen Center in San Francisco and for several years I was on the Zen fast track as buyer, guest manager, head of the meditation hall, head resident teacher, president, chairman of the board. During that time my marriage came apart, and eventually I dropped out again to work at Greens, our new vegetarian restaurant in San Francisco.

Since we had an excellent kitchen crew, I concentrated on working the front of the house: busboy, waiter, floor manager, host, and eventually wine buyer and co-manager. I left Greens at the end of 1983. *The Tassajara Recipe Book*, to which I contributed, came out in 1985, and then I had the good fortune to assist Deborah Madison, the founding chef at Greens, on her work *The Greens Cookbook*.

Ever since my days as a cook at Tassajara I have been interested in teaching others to cook. In more recent years I have taught cooking classes in the San Francisco Bay Area, nationally, and in Europe. I love this activity of teaching and encouraging people to cook. It's so nourishing for all of us.

READING THIS BOOK

This book is arranged so that stories or talks are followed by short recipe sections. You can read or cook, skip the recipes or skip the stories, follow the whimsy of the book or your own wandering.

Like many of us I have an aversion to straight lines. I don't want to be regimented as though I lack the capacity to sort things out for myself. In Taoism or Feng Shui, "evil" or "evil spirits" are said to travel in straight lines; what is evil does not know how to turn or bend. Similarly the present-day artist Hundert-wasser says that the straight line is the invention of fascists; he makes the floors of his buildings in Vienna uneven, so that the foot does not lose its intelligence. Despite all the idiosyncrasies of his buildings, they took no longer to construct because the work was more engaging for those involved.

Like cooking, reading this book can go in any direction you like.

THE RECIPES

Enjoyable everyday cuisine is the theme of the recipes collected here. This is not the same as a curriculum of great recipes you should know and master in order to impress people with your accomplishment, nor is it a complete or well-rounded selection of recipes from each of various categories. These are dishes I actually cook for myself, recipes I like and find myself coming back to.

When I cook I want something fairly simple and good – it doesn't have to be ingenious or magnificent, it needn't be the "best" or the "greatest." I start with decent ingredients and take good care of them.

And I am not concerned that you follow the recipes closely, because my real wish is that you actualize and deepen your own capacity to cook, that you be moved by the ingredients and respond accordingly. Please, you are welcome to surpass my humble endeavors with your own marvelous creations.

I began teaching cooking classes in the mid-eighties. My friend Susan Radelt invited me to lead some classes at her house in San Rafael. We had a wonderful time.

The next year Patty Unterman kindly offered to insert a notice about my cooking classes in the "Dining Out" column she wrote for years in the *San Francisco Examiner*, and hundreds of people responded. So for three months I held five classes a week, traveling with a cooler in my car: Tuesday night in Berkeley, Wednesday lunch in San Francisco, Thursday lunch in San Anselmo, Friday lunch in Sausalito, and Friday dinner back in San Francisco. Wasn't that something? A regular porta-kitchen.

At first I thought I should be teaching people a repertoire of dishes, like a curriculum. After several years I decided to teach "My Favorite Dishes" instead, and that is by and large what I have been doing ever since.

Where the recipes appear following a portion of text, I have often grouped two or three together as a possible meal. In other instances I have collected together recipes by category, be it potatoes, radishes, or desserts. You may also find the index useful for finding recipes of a particular category or those utilizing particular ingredients.

The Zen teacher Nan-Chuan said, "I tell my students to put themselves in the time before Buddha appeared in this world." Not just meditators, but also cooks need to be reminded that there is not a singular right way to cook. You have the capacity to sort things out for yourself, to find out how to cook according to your own sensibilities and experience.

So, please, carry on. I encourage you to keep finding out how to cook, how to live your life.

Cultivating Awareness

When You Wash the Rice, Wash the Rice

When I arrived in April 1967 to undertake my role as head cook of the newly founded Zen Mountain Center located at Tassajara Hot Springs, I soon became acquainted with the food habits and rituals of the residents. The center had not yet officially opened, but about twenty-five people were already living there. During my first meal preparation, someone informed me, "We do not use salt in the cooking."

I was stunned. I couldn't imagine such a thing. "You don't use salt?" I stammered. No, of course not. The custom was explained to me as though I was from another planet, as though it were the most obvious thing. "We don't use salt in the cooking because salt is bad for you. Everyone eats too much salt." This explanation didn't explain anything to me.

Arbitrary rulings are pretty common in community life everywhere. Someone knows what is right for everyone else, and although the rationale is vague and incoherent – conveying no real information – the authority wants you to go along with it (for your own good).

I found the idea of not using salt upsetting and disconcerting, but not being particularly adept at negotiation or inclined to throw my weight around, I went along with it until I had a chance to consult with Suzuki Roshi, our Zen teacher. These are, after all, the kinds of matters that can be easily resolved by higher spiritual authority.

"What shall I do?" I asked him. "Everybody has all these different ideas."

"Different ideas? Like what?"

"They don't want me to use salt. They say it's bad for you," I told him.

"You are the head cook," he said, "you can use salt if you want." The things a

Zen teacher has to clarify. I was relieved. I wanted everyone to be happy and to agree – but they didn't. I didn't want to side against anybody, but the Roshi's authority settled it for me. I could use salt.

Then I asked the Roshi if he had any advice for me as the cook. His answer was straightforward and down-to-earth: "When you wash the rice, wash the rice; when you cut the carrots, cut the carrots; when you stir the soup, stir the soup."

"OK," I decided, "I'll make those words my life." They became a life jacket, the proverbial Buddhist "raft," something which keeps you afloat, even when you are going under.

Some of my companions complained about missing meditation or lectures in order to prepare meals or to clean up after them. They seemed to think that Zen was happening somewhere else, and that we kitchen workers were missing out.

I would remind myself what our teacher had said, that work was also spiritual practice, another opportunity to see into the nature of things. I decided that I would prove it was true, that I would work as though it was indeed spiritual work. I didn't know any better.

So I worked hard. I worked at washing the rice when I washed the rice, cutting the carrots when I cut the carrots, scrubbing the pots when I scrubbed the pots. Complaints, fatigue, daydreaming, obsessive thinking – everything was met with a kind of admonition, a kind of reminder, "Just do it. Do what you are doing." I tried in a simple, direct, awkward way to be present, to see the rice with my eyes, to feel the rice with my hands, to have awareness in the movement of my arms. It certainly wasn't glamorous, and nobody said, "Why, thank you so much."

Day after day I put my awareness into activity, trying to find out how to cut vegetables, mop floors, clean sponges. I held the knife this way and that, trying out various cutting motions. Feeling my hands, I sought to use them more effectively and proficiently.

Overlooked details of activity would burst into my awareness. While intent

Cultivating Awareness

on cutting, I would hear the knife clatter carelessly onto the table. Not just cutting the carrot took attention, but picking the knife up, putting the knife down, wiping the knife, cleaning the knife, sharpening the knife, storing the knife.

I noticed that I needed to develop a more relaxed form of concentration. I would be concentrating so hard on cutting something that the smallest interruption –"Where's more salt?"–"What shall I do next?" would be shattering. My awareness would have to be more resilient than that, less brittle – focused, but willing to be interrupted and responsive to the next moment, whatever it turned out to be.

My awareness was also too narrowly focused within my body: While my hands and arms were actively engaged, my shoulders were aching with stiffness. I would have to let my awareness spread out, encompassing shoulders, back, stomach, hips, legs, feet, as well as hands and arms.

Anyone can do this kind of work. Whole worlds come alive. Entering into activity you find the world appears vivid with spinach, lettuces, and black beans; with cutting boards, baking pans, and sponges. You let go of the imagined and hypothetical so that awareness can function in the world of things. Where previously you may have hesitated or waited for the world to provide entertainment or solace, here you enter a world vibrant with the energy and devotion flowing out of your own being.

Food appears.

Recipes

Here's a menu where you can practice washing the rice, cutting the carrots, and stirring the soup. Since the dishes do not require elaborate preparations, you can concentrate on being at home in the kitchen and plunging into each activity.

Any time you cook is also an occasion to develop your taste and culinary sensibility by sampling as you go along. Often people follow a recipe, putting every-

thing in the pot, and then tasting it. This way you have missed the opportunity to learn what each ingredient contributes to the dish – why it tastes the way it does.

When you add one ingredient at a time, tasting before and after, you learn what flavors each ingredient provides or what elements in the dish that ingredient accentuates. Please, find out and know for yourself through your own experience.

Brown Rice
Lentil Soup with Cumin, Coriander, and Lemon
Ginger Asparagus

Brown Rice

Brown rice can be long grain or short grain, and I prefer the short-grain variety. Like wheat, potatoes, or other staples, rice is such a basic, wholesome food that simply cooking it well will bring out its natural goodness.

Japanese white rice is washed several times to remove talc, which is used as a preservative, while brown rice is washed more as a precautionary "you-don't-know-where-it's-been."

A pot with a tight-fitting lid is essential for this cooking method. If the lid is dented or the wrong size, excessive steam will escape while the rice is cooking, which means that the rice will need more liquid or more time to cook. If the lid does not fit well, cover the pot with foil, and put on a lid to hold the foil in place.

One cup of rice will serve 3 to 4 people

  1 cup brown rice
  2 cups water
  2 to 3 pinches salt

Rinse off the rice in a pot or bowl of water, and pour off the wash water. Place the rice, water, and salt in a heavy-bottomed saucepan. (Cooked in a pot with a thin bottom, the rice is likely to become brown and crispy on the bottom and can burn more easily, although you may be able to compensate for this by using a heat deflector.)

Cover the pot and bring to a boil over high heat – some steam will be escaping. Reduce the heat to its lowest setting, and let the rice cook undisturbed until done. This will take 40 to 45 minutes.

Not looking in the pot will allow the steam to go into the rice, instead of the air. Rather than observing the rice directly, you can observe the sound of the water bubbling in the pot, the small wisps of steam escaping, the changing aroma of the rice as it cooks.

The rice is done when the bubbling subsides, little steam continues to escape, and the aroma has blossomed. If you listen carefully to the pot, you can hear the distinct crackling or popping sound of the rice toasting on the bottom of the pot.

Turn off the heat, and let the rice sit for 5 to 10 minutes before serving. Uncover, and gently fluff rice with a fork.

Lentil Soup
with Cumin, Coriander, and Lemon

I made this soup at a cooking class recently, along with pimiento pizza and pear crisp. Several students thought it was the best part of the dinner. Lentils are like that – ordinary yet winsome. See what you think.

You can make the soup especially appealing if you grind the seasonings freshly (in a spice grinder or coffee mill), and use a good lemon (organic if possible).

This is also a fine soup for getting to know the ingredients by tasting carefully before and after adding each one: tasting the lentils, then lentils with vegetables and garlic, then with cumin, with coriander, with lemon, and finally with parsley.

Serves 4 to 6 people

1 cup lentils

8 cups water

Bay leaf

1 medium yellow onion, diced

2 tablespoons olive oil (optional)

2 cloves garlic, minced

2 stalks celery, diced

1 large carrot, diced

1 teaspoon cumin seed, freshly ground

2 teaspoons coriander seed, freshly ground

Peel of ½ lemon, minced

Salt (optional)

Flat-leaf parsley, a few sprigs minced, for garnish

Sort through the lentils for stones or other debris. Place in a large pot, add water and bay leaf, and bring to boil. Reduce heat and simmer 30 to 45 minutes, until the lentils are soft. The lentils could also be pressure-cooked. Once they are soft, see what they taste like.

If you want the soup to be ready soon, sauté the onion in the olive oil for several minutes, until it is translucent. Then add the garlic, celery, and carrot. Sauté a couple more minutes, then add a bit of water. Cover, reduce the heat, and cook until tender. Add to the cooked lentils. Season with the cumin, coriander, and lemon peel. Salt may be needed – if it isn't bad for you.

For a more leisurely soup, add the onion, garlic, celery, carrots, cumin, and coriander to the lentils after they are tender. Continue cooking 30 to 40 minutes, until the vegetables are soft. Add the lemon peel.

Before serving, check the seasoning and garnish with the parsley.

Ginger Asparagus

The question with this dish will be how to get that fresh ginger grated. There are small ceramic "ginger graters" that are moderately useful until, within a few seconds, the tiny ridges clog up – and then you need an old toothbrush to clean them.

What is really most fabulous and efficacious (and difficult to procure) is the traditional bamboo "washboard" ginger grater from China. You grate the ginger against the cross slates of bamboo, and, when finished, you just knock the grater against the counter to remove any grated ginger. A quick dip in water rinses off the grater.

Alternatively I use a regular cheese grater rather than one of the "ginger graters." Now, setting aside what kind of grater to use, here's the recipe. Of course if asparagus isn't available, pick something else that is in season.

Serves 2 to 4 people

1 pound asparagus
1 tablespoon olive oil
1 tablespoon (about ½ ounce *or* so) fresh ginger, grated
Salt
Touch of lemon juice *or* rice vinegar

Snap off the tough ends of the asparagus by hand. See if you can feel where the tough end snaps off pretty easily, indicating that the remaining stalk of asparagus is fairly tender.

Cut the asparagus diagonally into 2- to 3-inch-long pieces. Preheat a skillet. Add the olive oil, then the asparagus, and cook, stirring or tossing over high heat for a minute or two. Then add the ginger and salt, and continue cooking another couple of minutes. Taste to see how you like it.

To complete the cooking add a spoonful or two of water to the pot, cover, lower the heat, and cook until tender.

Season to taste with a touch of lemon juice or rice vinegar.

Careful Observation of the Obvious

Some years ago I called up a seed company to place an order. I wanted to get some red clover for a ground cover – the stalks of deep-red blossoms are quite splendid. I also wanted some garden netting to cover the seeds to keep the birds off. I ordered some other seeds as well to make up a minimum order and waited.

In a week or so a small box showed up. It didn't appear to be large enough to contain the pound of clover seed or the netting. I wondered how that little box could possibly contain the six by twelve feet of netting I had ordered, let alone all the seed. Another five or six weeks went by while I waited for my "real" order to come. The box sat on my kitchen counter until finally one day when I had nothing better to do, I decided to open it.

Inspecting each item, I discovered that everything I had ordered was in the box. The netting wadded up like a down sleeping bag into a compact bundle, and the pound of clover seed was just a fist-worth. Each seed packet also had a short message: "The Best Teacher of Gardening – Careful Observation of the Obvious." I had not even gotten out of the house into the garden yet, but the significance of the message was obvious. "Not only gardening," I thought, "but cooking too, and opening packages."

The obvious is what is easiest to overlook: Cookies on the bottom shelf of the oven burn sooner than the ones on the top. We think we know that salt is salty and sugar is sweet, but more than one cook has mistaken one white substance for the other. Food heated over time gets cooked, but only careful observation determines the moment it is at its peak of flavor and texture. Vine-ripened tomatoes eaten in the summer sun awaken warm surges of joy and vitality never encountered with tomatoes from the supermarket. What's important anyway? Careful observation can determine which differences make a difference.

At the New England Culinary Institute a young woman asked me for some zen secrets for dealing with stress. I explained that what I had noticed was that I would get more frantic, more speedy, and more stressed out in an effort to "catch up" with the things still to be done, and the things I must be forgetting. I wouldn't even take the time to check on what I might be forgetting because then I would be even further "behind."

Eventually, though, I noticed what was obvious about this situation. I could only do one thing at a given time. Getting hasty or volatile didn't help to get that one thing done. Also, worrying about what I still needed to do was a distraction from doing that one thing. I began making a checklist of what needed to be done, prioritizing it, and then doing the next thing, one after another. "That helped me," I said.

"That isn't zen," she said. "We do that all the time."

"Fine," I agreed, "just keep it up. There is nothing more zen than careful observation of the obvious. If you keep practicing that, you'll figure it out, whether it's how to cope with stress or what to do with a potato."

Garnishes

One thing I have observed is that garnishes often make the difference between an ordinary dish and one that is appetizing and flavorful. The garnish provides not only the final dash of color, but also some element of tart or pungent, earthy or vibrant flavor. Here are some initial notes that you can supplement through your own observation.

If you do nothing but garnish every dish, your cooking will change overnight.

FRESH GREEN HERBS

Parsley, thyme, lemon thyme, lime thyme, marjoram, oregano, mint, basil, cilantro, tarragon, rosemary – these are the ones I use the most. I include green onion or chives in this category as well. Aside from basil, cilantro, and tarragon, these grow in my garden most of the year, ready for use, a few steps out the door.

I find that fresh herbs are most aromatic and effective applied to a dish at the end, as a garnish, either minced or as whole sprigs. These are not just appetizing additions but are considered in Chinese medicine and Western herbal lore to be excellent digestive aids that work by counteracting the sluggishness that can follow eating. Even Peter Rabbit knew that: Being too full (and fat) to fit under the gate, he went back for some parsley to assuage his feeling of heaviness and indigestion.

I write more about herbs in "Root, Shoot, Flower, and Fruit" (page 239), but for now I would like to mention that I much prefer flat-leaf (Italian) parsley to the crinkly leaf varieties, as it is easier for me to mince, feels more pleasant in the mouth (when whole pieces are added to salads), and has (for me) a more complex and dynamic flavor. Consequently, I call for it in most all the recipes, but as my friend Kay said when she was trying out recipes, the crinkly kind works too.

PESTOS OR OTHER HERBAL MIXTURES

A classic pesto is made with fresh basil, Parmesan cheese, walnuts, olive oil, lemon, garlic, and salt. I use the term *pesto* to refer to any number of pureed herbal mixtures, which basically I make to taste, using fresh basil, cilantro, or tarragon. I also vary the pesto by using grated Asiago, choosing among a variety of ground nuts, and using lime or orange for the citrus. I may puree them or I may leave them chopped or minced. If I want a fresher, less oily flavor, I blend the herbs with orange juice and shallots, lemon peel, salt, and pepper.

CITRUS PEEL

Tartness is an important flavor component which often needs accentuating. Citrus peel can work well for this. I remove the colored part of the rind with a vegetable peeler, then cut it into thin strips, or mince (see page 63).

ROASTED CHOPPED NUTS OR SEEDS

I use primarily almonds, walnuts, and sesame seeds for this, but occasionally pine nuts, sunflower seeds, cashews, pecans, and hazelnuts. The nuts or seeds are roasted in a 350-degree oven for 8 minutes or in a skillet on top of the stove until they are fragrant, then sliced or chopped as necessary.

GRATED CHEESE

Though I love cheeses, I've cut back on my use of them in recent years, finding that I have better energy when my consumption is moderate. I still keep a block of Asiago cheese in my refrigerator, which I can grate freshly to provide a cheese garnish.

Grated cheese can be the white of Jack, Muenster, or Fontina; the orange of Cheddar; or the yellow of Gouda or Edam. It also provides depth or richness to a dish, with its high amounts of butterfat and its earthy flavor characteristics.

I also enjoy provolone because of its slightly smoky flavor, and Bruderbasel, which is the best smoked cheese I've found.

The last time my daughter Lichen returned from France she brought a dozen glorious cheeses, but that is not a matter of garnishes. Then I get some decent bread, some sweet butter, an everyday French red, and call it dinner.

OLIVES, CAPERS, SUN-DRIED TOMATOES

I tend to have these items around as they keep fairly well and can add color and strong flavor elements to a dish. There are so many fine olives these days, including Niçoise, Kalamata, and oil-cured. To pit the olives I smash them with the flat side of a knife to loosen the flesh before removing the pits. I find pizzas and pastas, especially, enhanced by these ingredients, but cauliflower, broccoli, and cabbage can also benefit. Keep them in mind. I get the dried rather than oil-packed sun-dried tomatoes and plump them in hot water before using. (See also page 17.)

Meeting the Zen Master

My friend Alan Winter and I decided to attend meditation at the Zen Center in San Francisco. We had talked about it a lot, but now we were actually going. A decision to do something, whether it is meditation or cooking, moves one's life along.

Inside the weathered but sturdy outer door of the Zen Center was a dimly lit foyer with broad stairs, a hallway, and an unassuming darkly hued door. Left to fend for ourselves, we took the stairs with its well-trafficked carpet.

At the top of the stairs we found some bookshelves, which turned out to be a shoe rack where one was expected to leave one's shoes and socks. The door to the left of the shoe rack was more formidable. It seemed to advise one to think carefully before entering, to prepare oneself and not just barge in. We paused while others came up the stairs behind us, removed their footwear, entered, and closed the door again. We could glimpse the interior with its golden light, its stillness and focus. On the other side of that door you couldn't fool around. What you did mattered. We finally entered.

Crossing the threshold we were in another world, a quiet, still room with a polished wood floor, mats and cushions in rows. People were already sitting cross-legged, facing the walls, and we found a couple of cushions on which to sit, facing the backs of others. We hadn't had any instruction, so we sat with our hands on our knees with the palms facing upward, because that's what we'd read about in the yoga books we'd looked at. Time passed.

Awareness came and went. I felt a lingering self-consciousness, a feeling of being watched, and I wondered if I was doing it right, the way it should be done. When I glanced cautiously around without turning my head, others seemed to be sitting quite intently. They seemed to know what they were doing, and they weren't noticing me. I stared at the floor. Nothing happened. My attention drifted.

A loud *whack* was startling, riveting. The whole room woke up. What was it? Sitting in the middle of the room, facing the backs of those sitting at the wall, I could observe that the Zen master was hitting people with a long stick of some sort. Again everyone seemed to know what they were doing. The master passed me by. The room returned to stillness. Seconds ticked by. A bell sounded.

After it ended, we tried to follow what others were doing: getting up, bowing, putting away cushions and mats. Then we had a religious service, which included bowing and chanting. Wonderful sounds moved through my being: harmonies, dissonances, chords, overtones, threads, pieces, the feeling of a fabric being woven, my body resonating, dissolving, being made whole.

When service was over, we lined up single-file to leave the meditation hall, exiting through another room, where the Zen master himself was bowing to each person in turn. I was meeting Suzuki Roshi for the first time, and I wondered what he would think of me, whether or not he would like me, whether or not he would approve of me. My turn came to bow.

I looked into his face, bowed, then looked into his face again. He seemed very ordinary. His face was impassive, with no trace of liking or disliking, approving or disapproving. What did he think of me? Not a clue. I felt vaguely disquieted or unsettled, yet along with his impassivity was an uncanny quality of openness. I felt received. What was going on? How unusual, how strange: I was put off by his nonresponsiveness, yet blessed with acceptance.

Many times we bowed like this after meditation, and I would never know what Suzuki Roshi was thinking, or if he was thinking at all. I was sure that he could "see right through me," and I was afraid he might not like what he saw. Yet even though he seemed to see completely through me, he didn't seem to be the slightest bit disturbed, annoyed, or upset.

Though his face showed no reaction I didn't sense that he was aloof or guarded or hiding what he must have been feeling, nor did I find him absent or somehow "not there." I had never met anyone like that – someone who seemed so completely present and receptive, yet unmoved on the surface, as though events reverberated and disappeared into some vast space. I felt grateful and privileged.

Years later, long after Suzuki Roshi's death, I came across a passage in Dogen

Zenji's *Shobogenzo* that reminded me of that first bow, that first meeting:

> What it is like is to be unstained.... Being unstained is like meeting a person for the first time and not considering what he looks like. Also it is like not wishing for more color or brightness when viewing flowers or the moon.

When we are looking at someone else, we commonly attribute or project our own thoughts onto the other person and determine what that person "must think of me," but I found this to be nearly impossible with the Roshi. Nothing seemed to stick to or to "stain" his awareness, so I could not construct or build up what his views must be. When I found myself burdened by what he must think of me, what he must want me to do, it would be patently obvious that he wasn't thinking any such thing. He was simply going about his business.

His presence, his awareness felt like such a great gift. Because he accepted himself, I was accepted, and because he accepted me, I could accept myself. What could be simpler? What could be more wonderful? Seen through and accepted.

Recipes

Looking at something clearly and carefully, we notice what its virtues are and how it might be used. We need a balance of flavors in our life; not everything is sweet. When we find out how to work with something, we may find it an excellent companion. Here are some things I've learned to use to help complement a range of dishes.

Red Onion Pickle
Sun-Dried Tomatoes
Working with Chilies

Cultivating Awareness

Red Onion Pickle

I still enjoy raw onion, but not in the large doses I used to, and over the years I've seen a lot of raw onion left on the side of otherwise empty plates. This has led me to appreciate the mildness of red onion pickle, which can be used for garnishing salads, adding to sandwiches, or enlivening vegetable dishes. They are relatively easy to make and convenient to have available.

For 1 onion

1 (6 to 8 ounce) red onion
Boiling water
½ cup red wine vinegar
Cold water
½ teaspoon salt
Some black pepper, Tabasco, or red chili, to taste

Cut the ends off the red onion and make a slit up one side, so you can peel back the papery skin and thinnest outer layer of onion. Cut a small piece off of one side so that the onion will sit flat, and then slice it thinly into rounds.

Place the sliced onion in a bowl and cover with boiling water, a cup or two. Let it sit for 5 to 10 seconds, then drain off the water.

Add the vinegar and a half cup or so of cold water to cover the onion, along with some salt and your choice of pepper. Let sit a half hour or longer. Drain to use. Or leave in the liquid and store in the refrigerator.

Sun-Dried Tomatoes

Sun-dried tomatoes have a wonderfully intense, concentrated flavor that is sunny, sweet, and robust, and their chewy texture is reminiscent of other dried fruits: a real taste buds pick-me-up. Although common in parts of southern Europe, they are relatively new to the grocery shelves of this country.

Sun-dried tomatoes come two ways: packed in olive oil and simply dried. The ones packed in olive oil are either plain or seasoned, most commonly with garlic and basil. Although I am quite fond of the oil-packed tomatoes, I usually buy the plain dried ones because they are a good deal cheaper.

The plain dried ones are too tough to chew, so I usually plump them in boiling water for 1 or 2 minutes, then drain, cool, and slice or chop.

Recently though, I came across a suggestion by Narsai David that worked well when I tried it out.

3 ounces sun-dried tomatoes
1 tablespoon vinegar
1 tablespoon wine or sherry
Olive oil
Fresh herbs *or* garlic

Combine tomatoes, vinegar, and wine in a jar with a tight-fitting lid. Set the jar aside for a day or two, inverting periodically. When the tomatoes become chewable, you can cover them with olive oil and herbs or garlic, to taste. Store them in the refrigerator.

Working with Chilies

Homemade Chili Powder (100% chili)
Homemade Chili Paste
Cutting Up Fresh Green Chilies

A whole new world opened up for me when I discovered chilies. Here's how to make your own ground chili, chili flakes, or chili paste.

I use three different dried chilies, which I get from one of the Mexican markets in San Francisco. Curiously enough chilies seem to be known by different names in different places, so I am providing a description with the name. If this is something that interests you, you can refer to Diana Kennedy's *The Cuisines of Mexico.*

CHILI NEGRO

This chili is about 6 inches long and dark red or black, turning a brownish tan when ground. The shape is cylindrical, about an inch across, and the surface is wrinkled. The flavor is distinctly earthy as well as hot.

ANCHO CHILI

Also known as *pasilla* in San Francisco, it is ruby red with a wrinkled skin and pear- or heart-shaped, 2 or 3 inches wide near the stem end and tapers to the far end. Ancho chilies generally range from 3 to 5 inches in length. The flavor has hints of prune and raisin. Ground, the chili is deep red, almost black.

NEW MEXICO CHILI

With its familiar brick-red color and supple, graceful length, this is the most recognizable dried chili. Generally about 6 inches long with a smooth-textured skin, this chili has sides that are roughly parallel, tapering toward the tip. The flavor is herbaceous, bell-peppery. When ground, the chili is an earthy red. These are called California chilies when they are grown there.

Homemade Chili Powder

Take a chili and grind it into a powder, that's the principle. Whereas commercial chili powders are most commonly made with garlic, oregano, and cumin in addition to chili, this is just chili. Then you can add the other ingredients in the amounts you wish, possibly to different parts of a dish, or to different dishes within the same meal. Also you may find you prefer a particular chili or combination of chilies.

People whose hands are sensitive find it best to wear thin rubber gloves when working with chilies. In any case, wash your hands thoroughly after handling the chilies, so you don't end up with a hot eye or some other body part touched by chili-laden fingers.

Yields about ¼ cup chili powder

1 ounce dried chili (about 5 chili negros, 3 ancho chilies, or 5 New Mexico chilies)

Roast the chilies on a baking sheet in a 350-degree oven for just 3 minutes. The chilies will tend to puff up a bit and become aromatic and a bit drier. More than 3 minutes and the chilies will tend to start burning.

Cool briefly, then cut open and remove the stem and seeds. Grind the flesh in a coffee mill, one you are using to grind spices and not coffee. Sometimes pre-chopping the chilies with a knife will be necessary so the grinder can do its work. That's it. I store the ground chili in small jars.

Note that ancho chilies are moister than the other two and will not grind as readily into a powder, but will end up in flakes. For most purposes this is sufficient, but if you want powder, you can roast the flakes in a dry skillet over moderate heat and then grind again.

An alternative method of preparing chili powder is to cut the stems off the chilies, chop them coarsely with a knife, and then grind them, seeds and all. After grinding, you can roast the powder in a dry skillet, stirring, for a toastier flavor.

Homemade Chili Paste

Making chili paste instead of ground chili powder is especially useful if you do not have a convenient way to grind the chili. It can be refrigerated or frozen for later use.

Makes 1 cup

4 to 6 dried New Mexico or negro chilies, or 3 to 4 ancho chiles (about 1 ounce)
2 cups water

Cut off the stems, cut open the chilies, and remove the seeds. Chop each chili into several pieces. Simmer in water over low heat for about 15 minutes. Puree – or scrape through sieve. Ready.

Cutting Up Fresh Green Chilies

I use mostly jalapeños and serranos. Again, great care is urged with the handling of fresh green chilies. These are even more potent than the dried chilies, so especially if your hands are sensitive, it's best to wear thin rubber gloves when working with them.

One way to chop up the chilies is to cut off the stem and then slice the chili crosswise. This will make healthy-sized chunks of chili, which can be quite formidable. My friend Dan Welch made me a pizza strewn with coarsely cut green chili, and I nearly died, while gamely assuring him that it was indeed (gasp) delicious.

A second method is to cut off the stem, cut open the chili lengthwise, and remove the seeds and pithy parts from the interior. Using a spoon for the seed removal will save your hands. Then chop or mince to desired consistency – I prefer rather finely chopped.

A third method, which I use, is to leave the stem on, and make lengthwise just-off-center slices to remove the flesh from the inner core of seeds. Then chop or mince the resulting pieces.

The Trouble with Thinking

During a week-long period of intensive meditation practice, I had my first opportunity to meet with Suzuki Roshi in a formal interview. I was making every effort to practice meditation the way I thought it was supposed to be practiced. I wanted something to show for my effort. Perhaps I could attain a state of "not-thinking" or a "calm mind." Perhaps I could attain "true realization." These sorts of attainments would certainly be better than making a lot of money or gaining other kinds of success or fame, wouldn't they? Well, I thought so.

The problem was I wasn't getting anywhere. Try as I might to concentrate on my breathing, I found I was almost constantly engaged in thinking: planning, remembering, evaluating, assessing, a perpetual sorting out of how I am doing, where I am now, where I need to get to, how to get there.

So when I went to speak with Suzuki Roshi I did not have any "thing" to show for my efforts. I felt humbled and somewhat frustrated and discouraged. What would the master think of this poor excuse for a zen student? I wanted him to like me, but I didn't see how he could. I certainly didn't. There wasn't much to like as far as I could tell. I entered his cabin and performed the required half-prostration bows, not directly to him, but toward the altar with its Buddha image, candle, and burning incense.

I bowed forward, kneeled, touched my head and forearms to the floor, and then raised my hands, palms upward. He corrected the way I was positioning my hands during the prostration: "When you lift your hands from the floor, hold them flat," he explained, gesturing, "as though you were lifting the feet of the Buddha. When you cup them like that, it feels as though you are trying to grasp something and being greedy." His voice was pleasant and matter-of-fact.

OK, I thought, I can do that. It was a relief to have something to work on, something to keep in mind, and he had shown an interest in my practice! Still I felt flustered and not particularly comfortable in his presence.

Then I sat down on the cushion opposite him, crossed my legs, and adjusted my posture. I didn't know what to expect, or what was expected of me, so I just sat there quietly facing him. The world turned. I don't think he had the slightest thought about my attainment or lack thereof. He seemed contained, quiet, and alert in repose. I began to relax. Finally after a few minutes, he inquired, "How's your meditation?"

"Not so good," I replied.

"What's not so good?" he asked.

"I can't stop thinking," I lamented.

"Is there some problem with thinking?" he questioned, and right at that moment when I looked directly for the problem, I couldn't actually find it. I felt relieved and lightened, but I wasn't ready to admit I couldn't find the problem. Besides, didn't he and the other teachers keep instructing us to follow the breath rather than think?

"When you sit zazen, you are not supposed to think," I explained.

"It's pretty normal to think," he stated, "don't you think?" His way of speaking was so innocent of attack: not contradicting, not belittling, not finding fault.

I had to admit that thinking was pretty normal, "but we're not supposed to think, are we?"

"The nature of mind is to think," Roshi explained. "The point of our practice is to not be caught by our thinking. If you continue to practice, your thinking will naturally change. Sometimes it will stop. Your thinking will take care of itself."

Reassured I continued to sit quietly, again waiting to see what would happen. The room was still and peaceful. After awhile the Roshi's voice was there again. "What is it you want most of all?" he asked.

A word came to me instantly, but I hesitated, and stopped to think it over. Was it really the answer, really what I wanted? Was it right? Was it good enough? Nothing else came to mind, so at last I voiced it: "the Truth."

I felt awkward saying it, uneasy admitting it, but there it was, "the Truth."

Yet the Roshi's silence, the silence of the room swallowed it up. There was a lot of room to grow in that silence.

After a while he said, "Please continue your practice." We bowed to each other. The interview was over.

I left feeling relieved, overjoyed, and eager to go back to sitting meditation. I was relieved that my teacher had not criticized me or condemned me for what I viewed as my poor practice; overjoyed that he actually seemed to appreciate my efforts; eager and willing to return to the meditation hall and try some more.

Meals I prepare do not always come out as well as I would like, and I can be critical or judgmental, but also I know not to worry too much about my thinking. I can appreciate my ongoing effort and return to the kitchen. My wish to nourish and feed myself and others sustains me.

If you have a similar wish to offer sustenance, I hope you will find ways to act on it. What you think one way or another about being a good cook or a bad one is of little concern. Once you acknowledge it, your wish will fulfill itself.

Recipes

When I think about desserts I usually think first of fruit crisps. They were my favorite dessert at YMCA camp, and years later I was delighted to find out how to make them. I love the buttery sweetness and juicy vitality.

These days many people think that butter and sugar are unhealthy, but I still believe in old-fashioned desserts. Thoughts can rigidify into hard and fast rules, creating a regimen that doesn't leave much room for enjoyment and satisfaction. I would rather be flexible and keep finding out for myself what truly nourishes me and what doesn't. Returning to my own experience and learning from it how various foods affect me is an engaging and absorbing activity, whereas trying to impose someone else's thinking or scientific findings on myself ends up being stultifying and at times demeaning.

*I don't have dessert that often, but when I do, these crisps bring me great plea-
sure, and I feel warm and thankful.*

Blueberry Crisp
Peach or Nectarine Crisp
Rose-Scented Sugar

Blueberry Crisp

*Fresh, ripe berries in the summertime are one of nature's remarkable blessings:
plump and succulent, with intense, concentrated flavors – a soothing sweetness
combined with a bracing tartness. I'm less interested in berries out of season,
shipped halfway around the world or canned or frozen.*

*Often I have fresh berries for breakfast or dessert with a sprinkling of sugar and
perhaps a few drops of balsamic vinegar, but they make for a quickly prepared
crisp as well. Here's the story.*

Serves 4 people

12 ounces fresh berries, such as blueberries, blackberries, ollaliberries, raspberries
2 tablespoons maple syrup *or* honey *or* sugar
2 tablespoons water
⅔ cup unbleached white flour
⅓ cup white sugar *or* Rose-Scented Sugar (see recipe page 27)
Small pinch of salt
⅓ cup unsalted (sweet) butter

Preheat oven to 375 degrees.

Place berries in an 8- or 9-inch round baking dish or ceramic casserole with
the maple syrup and water. I use a touch of water here for added juiciness.
Combine flour, sugar, and salt, and then cut in the butter with 2 knives or a

pastry cutter. Distribute the topping over the berries and bake for 35 to 40 minutes, until the top is browned and the juices from the berries bubble up.

Peach or Nectarine Crisp

Peaches in season are a luscious and sensuous treat, but nectarines make a good substitute.

Serves 6 to 8 people

4 to 6 peaches (or nectarines), depending on size
Juice and grated peel of 1 lemon
1 teaspoon cinnamon
½ teaspoon freshly grated nutmeg *or* ¼ teaspoon mace
⅔ cup brown sugar
1 cup whole-wheat flour
2 to 3 pinches of salt
½ cup unsalted (sweet) butter

whipped cream *or* ice cream (optional)

Preheat oven to 375 degrees.

Put the peaches in boiling water for 10 to 30 seconds to loosen the skins, so that they may be peeled easily – the nectarines do not need peeling. Remove from the water, drain, peel, and cut into slices. Toss with the lemon juice and peel, cinnamon, and nutmeg, and arrange in a 9- by 13-inch baking pan.

Combine the sugar with the flour and salt and cut in the butter with a pastry cutter or 2 knives. Distribute the topping over the peaches. Bake for 35 to 40 minutes at 375 degrees or until the peaches are fork-tender. Serve warm with a dollop of whipped cream or small scoop of ice cream if desired.

Peach Crisp: I also like the peach crisp with 1 teaspoon allspice in place of the cinnamon and nutmeg.

Pear Crisp: Use 4 to 6 pears in place of the peaches, and ¼ teaspoon of freshly ground cardamom in place of the nutmeg.

Apple Crisp: Use 4 to 6 pippen apples in place of the peaches with the same combination of spices.

Rose-Scented Sugar

An excellent "secret" addition to fresh fruit desserts. One of the roses in my garden is especially fragrant, and the aroma of the rose is what makes the sugar what it is.

I first came across this item at some cooking classes with the master chefs of Szechwan. I don't know how they make it, but this is what I do.

Makes 4 to 6 cups sugar

3 to 4 roses
4 to 6 cups sugar

Find some aromatic roses. Remove the petals. Layer the petals in a jar with the sugar. Cover.

Leave covered for 2 to 3 days or longer. Check the aroma. The sugar will absorb much of the moisture in the fresh flower petals. Then, if the jar is kept closed, the flower petals will begin to rot. So after a few days, when the sugar is wet, you have a choice: Remove all the flower petals, or leave the lid off the jar for some days to let the sugar dry out a bit. Then it can be left covered and the petals removed later.

Acting on Your Own Recognizance

Who says you can't cook? I give you permission. You can look with your eyes and feel with your hands, smell with your nose and taste with your tongue. You can think and create, be inspired, or stumble along. You keep finding your way.

Some months back a couple of friends thought they would make a series of videos of me teaching cooking classes and talking about Zen this and that. However, when we did a sample video, I came across rather woodenly, so we shelved the idea. Turns out I am camera shy. In the meantime my would-be producers had written a fund-raising letter promising, "Ed Brown will teach even inveterate meat eaters how to produce vegetarian masterpieces." I begged to differ.

"Excuse me," I said, "but that's just the point. I am not going to teach people how to produce vegetarian masterpieces, as though there were no reason to cook unless they could make a masterpiece. I want to encourage people simply to cook, to be willing to cook ordinary food they find enjoyable to eat. I'm trying to remove the pressure people feel to produce masterpieces or 'don't even bother.'" Since the project was shelved, we didn't need to rewrite the letter.

You can learn many things about "cooking," about ingredients, cutting, combinations, and procedures, but even more fundamentally you can learn to act on your own experience, outside of recipes, relying on your innate capacity to taste and sense and decide for yourself what you like. By this I do not mean following your "instincts," which seems to me a rather amorphous concept, but being present, carefully observing the obvious, acquainting your palate with your palette.

When I helped Deborah Madison write *The Greens Cookbook* we worked very hard to produce a well-crafted manuscript. She edited the recipes on which I worked, and I edited her recipes. Then the two of us went through all the material together, and finally we went through the whole manuscript with a cookbook editor, checking everything: Does Parmesan have a capital "P" every time it is used? Is that "4" or "four"? We thought we had a highly polished draft, so we were dismayed when the manuscript came back with numerous pink press-apply labels sticking out the right side.

Where we had written, "Cook the onions until they are translucent," the little label would read, "How long?" Where we had written, "Season to taste with vinegar," the question was, "How much?" Deborah and I were pretty frustrated and annoyed because we were trying to teach people to COOK!, not by following directives but by paying attention to the process. We were giving out visual and sensory cues, not times and amounts. Are you going to cook by looking at the food or by looking at the clock?

Finally we came to a recipe in the pasta section where we had written, "Cook the vegetables until they are as tender as you like," and our editor asked, "How long? How do we know?" We threw up our hands. "If you don't know what you like, who does?" we raged at the heavens, or "Establish a standardized 'chew' which you will use to test whether or not something is 'tender,' then place food in mouth and apply standardized 'chew.' If standardized chew manages to divide food in mouth, call that 'tender,' say you 'like it' by definition."

But clearly there is no definitive answer. You just have to wing it and feel for yourself. You're the expert on whether or not you like something. You have eyes and ears, a nose and a mouth, likes and dislikes (which can be revised sometimes). You can learn to trust your own taste, which will change and develop, get tired or be stimulated, as you go along.

Some years ago *The Wall Street Journal* printed a couple of articles about food. The first, called "Even Canned Corn Stumps Modern Cooks," said that the Pillsbury Company had tried taking the directions off of its canned corn but got so many calls from consumers wondering what to do, that it put the directions back on. The directions read: "Put corn in saucepan on heated burner."

I thought that was pretty funny, and I was laughing when I told the story to a friend of mine, forgetting that her husband does most of the cooking in their house. "But Ed," she protested, "I'm like that. Do you drain the corn or not?"

She is an incredibly creative and productive artist and an amazingly good-hearted friend, and I cannot believe that she is as clueless as she makes it sound, but when it comes to something as esoteric as canned corn, you better hope her husband is cooking.

Accompanying the canned-corn article was one headlined "How Much Will People Pay Not to Cook? Plenty." This piece pointed out that people will pay three to five times as much money for prepared foods as for the plain ingredients, even for something quite simple like scalloped potatoes. This basic fact explains why turning raw materials into "product" (being a manu-facturer) is a much more lucrative business than producing the raw materials (being a farmer).

Personally I do not understand how people can afford all these "value-added" products. Maybe I just don't work hard enough at earning money, but I find the work of handling the ingredients themselves to be quite satisfying. This is why I am so pleased when I receive a letter from someone who senses the liberation in learning a basic cooking skill. One reader wrote to tell me that he felt as though he had completed a rite of passage: "Since I have learned to make bread from your book," he said, "I feel as though I have re-owned my life from corporate America."

We all have the capacity to re-own our life. We are wonderfully capable of finding out how to live in a way which is nourishing and satisfying. Thank you for your effort.

Recipes

Making pizza is a labor of love. I hope you will enjoy it too, as it is such an engaging and satisfying activity, and the results are so crowd-pleasing. I never thought much about making it until I went to work at Greens Restaurant and began eating our pizza almost every day. Later, when I worked with Deborah Madison on The Greens Cookbook, *I helped to develop the pizza recipes. What follows is my version of a pizza I had more than once at a restaurant in Venice. I believe it is an excellent way to introduce you to the basics of my current style of making vegetarian pizza, which is elaborated in "Pizza Making Ruminations" following the recipe.*

Pizza Venezia

Pizza Venezia

Usually a pizza has its ingredients distributed evenly over the surface, so that each slice is the same, but this pizza has a variety of toppings each of which is put in its own area, making it something of a work of art. The arrangement is not supposed to be done with straight lines but rather with whimsy: Roasted red peppers conjoin with mushrooms or asparagus; carrots are flanked by zucchini or eggplant. When the pizza is sliced, each piece is different.

Since people's appetite for pizza varies enormously, you will have to decide for yourself how many people a 12-inch pizza will serve.

Makes 1 (12-inch) pizza

1 batch Pizza Dough (see recipe page 34)

4 cloves garlic, minced
5 tablespoons olive oil
1 Japanese eggplant (4 to 6 ounces)
1 zucchini (4 to 6 ounces)
1 Roasted Red Pepper (see recipe page 81)
6 to 8 ounces mushrooms
Salt
Black pepper
8 ounces asparagus

1 teaspoon fresh ginger, grated

1 carrot (about 4 ounces)

2 teaspoons horseradish

½ cup green onion, sliced (white and green parts)

Balsamic vinegar

Lemon juice or rice vinegar

2 shallots, sliced

4 ounces cheese, grated, such as Jack, pepper Jack, Gouda, provolone, mozzarella, Muenster

Optional:

8 to 12 cherry tomatoes, cut in half

6 kalamata olives, pitted and chopped

1 to 2 teaspoons capers, drained

Garnish:

2 teaspoons grated lemon peel

1 tablespoon fresh thyme, minced

Preheat the oven to 400 degrees.

This recipe involves making six different toppings, but you are certainly welcome to see what you have available in the way of leftovers to get you started, or make fewer toppings but *more* of the ones you do make. See "Pizza-Making Ruminations" below for general advice on preparing pizza.

Start the Pizza Dough (see recipe page 34), and set it aside to rise.

Combine the minced garlic with 3 tablespoons of the olive oil. Cut off the ends of the eggplant and slice it in diagonal pieces between an ⅛ and ¼ inch thick and 3 inches long. Cut the zucchini the same way. Place the slices on an oiled baking sheet and brush the tops with olive oil as well. Bake in a 400-degree oven 15 to 20 minutes, until the zucchini is tender, and remove it. Continue baking 15 or 20 minutes, until the eggplant is tender. Set vegetables aside in separate bowls. Turn up the oven to 525 degrees.

Cultivating Awareness

Prepare the Roasted Red Pepper (see recipe page 81).

Slice the mushrooms and sauté them in a tablespoon of the olive oil for a few minutes, then add a spoonful of the minced garlic (from the garlic and oil mix) and a touch of salt. Continue cooking several minutes longer, until the mushrooms are well browned. If there is extra liquid, remove the mushrooms and cook the liquid down until it is syrupy, then add it back to the mushrooms. Season with pepper.

Snap the tough ends off the asparagus, and then cook it, whole or sliced. My idea here is the Ginger Asparagus (page 9).

Cut the carrot diagonally into ovals. If you enjoy cutting, slice the ovals into strips. Sauté the carrot in 1 to 2 teaspoons olive oil over high heat for a couple of minutes, then add the horseradish, green onion, and a pinch of salt. Cook another minute or two, then cover, turn off the heat, and let the carrot steam.

Before assembling the pizza, do some seasoning: Sprinkle the eggplant with about 2 teaspoons of balsamic vinegar, some salt, and pepper. Sprinkle the zucchini with a touch of lemon juice, if it's handy, or some rice vinegar, salt, and pepper.

You can leave the roasted red pepper in big pieces or cut it into slices. Sprinkle with salt and pepper, a teaspoon of the garlic oil, and a touch of balsamic vinegar.

Roll out the pizza dough into a circle about 12 inches in diameter and about ⅛ inch thick. Place on a pizza baking pan of the same size. Since I do not like to see people discarding pieces of my crust, I prefer not to make a rim of dough at the edge of the pizza. Instead I spread the toppings all the way to the outer edge.

First spread out the remaining olive oil with the garlic. Then distribute the sliced shallots and the grated cheese. Now put each topping in a particular area on the pizza. Notice which colors might go well together. If you are at a loss, put the toppings in pie-wedge areas or in strips across the pizza.

If using the tomatoes, olives, and capers: Top the eggplant and zucchini with the cherry tomato halves. Top the carrots with the chopped olives. Sprinkle the capers where you want them, perhaps with the asparagus and/or the zucchini.

Ready to bake! Bake on the top shelf of a very hot, 525-degree oven for 12 to 15 minutes, until the bottom of the crust is browned.

Remove pizza from the oven and garnish with the lemon peel and fresh thyme. Since I do not have a pizza cutter, I slide the pizza off the pan onto the counter or a cutting board and cut it into slices. Then I put the slices on a platter to serve.

PIZZA DOUGH
> *For 1 (12-inch) pizza*

½ cup warm water

1 tablespoon active dry baking yeast (1½ packages)

¼ teaspoon honey

2 tablespoons olive oil

½ teaspoon salt

½ cup whole-wheat flour

½ to ⅝ cup unbleached white flour

Make sure the water is warm, but not hot – just about body temperature, so it doesn't even feel warm on your hand. Add the yeast, honey, olive oil, salt, and the whole-wheat flour. Beat well with a spoon, then stir in ⅜ cup of the white flour, and use the other quarter cup of white flour, as needed, to knead the dough for several minutes until it is smooth and elastic. More complete instructions for working with a yeasted dough are included in White Bread with Cornmeal, beginning on page 44.

Oil the bowl and put the dough back in. Set it aside and let it rise for 25 to 30 minutes while you prepare the other ingredients for the pizza. If you are not yet ready for the dough, punch it down and let it rise again, up to another 20 to 25 minutes.

The first time I offered to make pizza for my companion Patti, she said that pizza often gave her heartburn and she didn't care for it that much anyway. I replied that she hadn't yet had my pizza. Soon she was a convert.

Making pizza can be quite enjoyable, partly because of the creative potential. Here are some of the possibilities and some of the "secrets":

- Precook most vegetable toppings. When baked, the pizza crust will brown and the cheese will melt without the vegetables on top cooking much.

  Precooking is especially important if the vegetable is, say, eggplant, which is one of the few vegetables I prefer cooked to complete softness – doesn't everyone? I say this because I was served a pizza with almost raw eggplant at one of those chic restaurants that all the critics fawn over. After attempting to remedy the situation, our trying-to-be-helpful waitress returned to our table chagrined and said, "The cooks say that's the way the dish is made." Well, not in my kitchen. (And we haven't returned to that stylish restaurant either.)

  Also some toppings, notably mushrooms, can release a fair amount of liquid during baking if they are put on raw, which can make the pizza too wet.

  Cooking the vegetables also provides an opportunity to season them, so these pizza toppings could just as well be a vegetable dish and vice versa.

  A few vegetables may be put on without cooking: fresh tomatoes, and, sometimes (if the audience doesn't mind), bell peppers or onion.

- Use some intense flavor elements to give the pizza some pizzazz: garlic, olives, capers, roasted red peppers, green chilies, red chili, ginger, sun-dried tomatoes, thinly sliced lemon or lemon peel, fresh and dried herbs. These can be mixed in with the vegetables or strewn over the surface of the pizza.

- Garnish the pizza *after* baking. This is especially important. Then the aromatic flavors of freshly grated Parmesan or Asiago cheese and fresh herbs are enhanced by the heat. If the cheese and herbs are put on earlier, then much of their flavor is left in the oven.

- Use flavorful cheeses (provolone, smoked, Gouda, Jack, pepper Jack, Muenster, goat cheese – the list could go on and on) in place of the standard, generic mozzarella. For my taste, it makes pizzas much more interesting.

  I read in *The Wall Street Journal* that Poland is taking to pizza, but one problem they have is that there is a shortage of mozzarella cheese, so they improvise, using cheese that is available. Apparently the Italians sometimes criticize them, saying that they do not understand the concept that one function of the cheese is to be gooey and hold the toppings in place.

  I haven't found this to be a problem, except when I do not use any cheese. Also I tend to use a modest amount of cheese on my pizzas, because otherwise there can be an extravagant amount of oily goo in any particular bite.

- Here are some pointers about making the dough and rolling it out.

  I now make a fairly basic olive oil bread for my pizza dough. At Greens we used to use half water and half milk for the dough and include some rye flour as well as wheat, so if you are so inclined, you can experiment to find out what you like. For instance, some cooks like to add a small amount of cornmeal to the dough or to sprinkle it on the pizza pan before putting the dough on it.

  I am fussy about rolling out the dough, because I am not a fan of "cardboard" crusts. As far as I can tell, pizza doughs are made with basically good ingredients, and become dry and pasty on the palate, because they are rolled out on an excessively thick mound of white flour. A perfectly fine dough now has a thick coating of dry flour on it, which will not be appetizing.

  I encourage people to "dust" the counter with flour by tossing a small handful of flour up (not down) and out across the surface. First shape the dough into a ball, then dust the counter and position the dough. Flatten out the ball by hand to start with, and then turn it over, so that there is a thin coating of flour on top.

  Now roll out the dough, working the rolling pin from the center to the edge in various directions. If the dough starts to stick, stop rolling, pick up the dough, and "dust" the counter again. Turn the dough over when you put it down again, and recommence rolling. Picking up the dough gives it a chance to "relax," making it easier to roll out further.

When the dough is repositioned after a fresh dusting, this is also a good time to reshape it by hand into something more closely resembling a circle (if that has started to be a problem).

Once the dough is the size of the pizza pan, pick it up and place it on the pan, which doesn't need greasing.

- Put the toppings out to the very edge of the pizza so there is no "crust" for people to discard. This makes the pizza feel especially ample and generous. If you make the pizza in this style, you do not need to do the showy thing of tossing the pizza in the air. (Oh, go ahead.) The dough is tossed after it is rolled out in order to stretch it thinner in the middle of the circle while leaving it thicker at the edge. Between stretches the circle of dough is tossed in the air to rotate it.

- Use a pizza stone if you own one. I don't have one of these anymore, and I seem to get by without it. These stones work best if they are preheated in the oven, which means you will also need a pizza "peel," which is like a wooden pizza "shovel." The pizza is assembled on the peel and, yes, with loads of flour underneath, so that when ready it can be slid off the peel onto the hot stone. That's a trick right there! At its best this makes a wonderfully toasty, nutty crust.

Finding Out How

Homemade Bread Touches My Heart

In the summer of 1955, when I was ten years old, my brother Dwite and I went to visit my Aunt Alice in Falls Church, Virginia. We flew first to Kansas City, and minutes before we landed I threw up, making use of one of those little bags which I had been naively asking about a short time earlier. Airplane rides were a lot bumpier then.

We sat in the plane on the tarmac there in Kansas City, waiting for something to be fixed, sweating in the terribly hot and stuffy confines. Isn't traveling fun? A bedraggled little boy arrived in Washington, DC, but was quickly revived by rest and old-fashioned hospitality.

Best of all was the homemade bread my Aunt Alice baked. I couldn't believe how good it was. We'd have it with dinner, and then in the morning we would toast it for breakfast. It was fabulously delicious, especially with the real butter and homemade jams we got to put on it. After the store-bought bread and margarine at home, it was simply to die for. Returning from a day of sightseeing in DC, we would be greeted with the hearty, earthy, nutty aromas of freshly baked bread.

What I could not understand was why more people were not baking bread at home, delighting their noses and pleasing their palates. In the stores then, pretty much all you could buy was foamy white bread. When allowed to, I would eat off the crusts, and then mash the rest of the slice into a marble. Then I had something solid to chew on. Commercial whole-wheat breads, if you could find them at all, were dry and comparatively tasteless.

Once I tasted my Aunt Alice's bread I wondered why people put up with the more boring version when they could be having bread which stimulated and awakened previously unknown reservoirs of joy and delight. Well, my thinking may not have been that sophisticated, but I decided then and there that I would learn how to make bread and that I'd teach others how to make bread. Plus, when I could, I'd get butter to put on it.

When my brother and I returned from our trip, I asked my mom if she could teach me how to make bread. "No," she said, "yeast makes me nervous." The directions in the cookbooks instructed one to "knead" the dough. No, Mom could not show me how to do that either.

Arriving at Tassajara eleven summers later, I encountered fragrant, satisfying bread and two chefs, Jimmie and Ray, who could teach me how to make it. When I asked if they would show me how, they were more than happy to do so. It felt like an initiation. The secrets of how to do something were being shared and passed on. I became a descendent in the lineage of bread-bakers.

Jimmie and Ray, it turns out, had learned to bake bread from Alan Hooker at the Ranch House Restaurant in Ojai, California. Years later, when I was working as a waiter at Greens, Mr. Hooker came to dinner. Afterward I went over to his table, introduced myself, and gave him autographed copies of my books.

"You don't know it," I said, "but I'm your disciple." Even though we had never met, I felt very close to him. We shared a deep love for wholesome bread, and he had learned this craft, worked at it, and transmitted it, opening a whole new world to me. I felt profoundly grateful and honored to be in his company.

Later I received a letter from Mr. Hooker, thanking me for the books and inviting me to have dinner at his restaurant if I was ever in Ojai. Fortunately, I was able to take him up on his offer, and we had a delightful time. A shared love of bread and baking brought us together.

This is culture, the passing on of how to do something. The shared know-how bridges the generations and gives life to life. Nowadays it is less obvious what we are passing on other than how to watch television and walk a supermarket aisle, and what is lost is not just the way to bake bread, but the connectedness with our predecessors, our fellow beings, and the stuff which is our life.

Recently my brother returned from a Fourth-of-July visit to Washington, DC, where his oldest son is living with his wife and baby girl. So I felt it was timely to remind Dwite of our earlier trip, when he watched after me and we discovered bread and took in the sights together. "What I remember is the Smithfield ham," he mused, "but it didn't change my life."

The whys and wherefores of hearts being opened is mysterious and momentous, giving shape to whole lifetimes.

Recipes

Learning to make bread from Jimmie and Ray, I became acquainted with the "sponge method." This method – where about half the flour is added initially, and this thick batter is allowed to rise – makes adding the remaining flour and kneading much easier than it would be otherwise, and the bread itself has a fine even "crumb."

If you are particularly interested in bread-making or would like more details and illustrations of the various procedures involved, you might enjoy my Tassajara Bread Book, *which has now been reprinted in a special 25th anniversary edition.*

Meanwhile, here are two of my favorite breads.

White Bread with Cornmeal
Overnight Wheat Bread

White Bread with Cornmeal

Once I started making bread, I began experimenting with putting in rye flour, barley flour, cornmeal, millet meal, and oatmeal to see what happened and how I liked the bread. I love the sunny color and flavor of this bread.

Makes 2 loaves

3 cups lukewarm water (90 to 110 degrees)

2 tablespoons active dry baking yeast (3 packages)

¼ cup honey

1 cup dry milk

1 cup whole-wheat flour

3 cups unbleached white flour

4 teaspoons salt

¼ cup corn oil

3 cups cornmeal

Up to 2 cups additional unbleached white flour

1 egg

2 tablespoons water

2 tablespoons poppy seed, if available

Put the lukewarm water in a medium-large mixing bowl. If the bowl is ceramic, preheat it first with a few cups of hot water. The temperature of the water is important, because the yeast that makes the bread rise will be killed if the temperature rises much over 125 degrees. Conversely, when the water is cooler, it simply means the bread will rise more slowly. Having the water just about body temperature ensures that the yeast will work well. Yeast, by the way, is about ten times cheaper in bulk at the natural food store than in those little packets.

Stir the yeast into the water, then the honey and dry milk, then the cup of whole-wheat flour. Don't worry about lumps because they will disappear in the process. Mix in the 3 cups of unbleached white flour to form a thick batter, and then beat well with a spoon, about 100 strokes. (To "beat" with a spoon means to make small circular strokes just in and out of the surface of the batter.) This will help to make the dough elastic. Set aside and let rise 40 minutes.

Fold in the salt and oil. Fold in the cornmeal a half cup at a time. (To "fold" means to scrape the spoon along the side or bottom of the bowl, and fold the wet dough there over the dry dough on top.) If you turn the bowl a quarter turn between folds, you will be approximating the action of kneading.

After adding the cornmeal, begin folding in the additional white flour, a half cup at a time, until the dough comes away from the sides and bottom of the bowl. Turn the dough out onto a floured board.

An aside here about "process": It's an efficient use of time and effort at this point to scrape the sides and bottom of the bowl as well as the spoon, and incorporate the scrapings into the dough. At Tassajara we used to have those yellow and red Tuffy scrub balls, and often people cleaning a bread bowl would gum up a Tuffy pad or a sponge with dough, since they had not done the scraping earlier. Yuck! Do the scraping now rather than discard a sponge later.

Knead the dough on a floured board, sprinkling on more flour, as needed, to keep the dough from sticking to the board. Knead for about 6 to 10 minutes, until the dough is smooth. (To "knead" means to lift the far side of the dough and fold it in half toward you, then to push down and away on the dough with the heels of your hands – the dough rolls forward and the top fold ends up about two-thirds of the way to the far side – then turn the dough a quarter turn, usually clockwise, and repeat the procedure.) Kneading is complete when the surface of the dough is smooth and feels like a baby's bottom.

Oil a bowl – if you scraped out the mixing bowl, you can use that one – place the dough in the bowl, and turn once so that the top of the dough is oiled. Let the dough rise for about 50 minutes, until it is doubled in size.

Punch down dough. (To "punch down" is to push your fists into the dough numerous times so that it is thoroughly deflated, although it will not be as small as it was originally.) Let rise a second time for perhaps 40 minutes, until it is doubled in size.

Divide dough in half, shape into 2 loaves, and place in oiled bread pans – I like the 4½- by 8½-inch pans. The simplest way to shape loaves is to flatten out the dough into a rectangle a bit longer than your bread pan, then roll the dough into a log shape, pinch together any loose ends, and squeeze into the pan.

Preheat the oven to 375 degrees. While the oven is heating up, let the bread rise in the loaf pans about 20 minutes or so, until again doubled in size.

Mix together the egg and water and brush it on top of the loaves. (You will have more than enough.) Then sprinkle on the poppy seeds.

Bake in a 375-degree oven on the middle shelf between 45 and 55 minutes, or until the tops, sides, and bottom are golden brown. Depending on your oven, you may want to move the bread from one shelf to another during the baking. On the lowest shelf the bread will tend to get overdone on the bottom, while on the topmost the bread will tend to brown on top.

Remove from the pans and let cool.

Overnight Wheat Bread (Wheat Veneration)

I love the intense pure wheat flavor of this bread, which is not weakened with the addition of milk or sweetener. The flavor of the wheat "blossoms" by letting the dough sit overnight. The bread is reminiscent of sourdough, but you needn't get involved with trying to locate or make a sourdough starter. For people like me, who are unfortunately not as devoted to baking as we once were, this bread, hearty and fulfilling, makes a satisfying substitute.

Makes 2 loaves

Evening:
¼ teaspoon active dry baking yeast
3 cups whole-wheat flour
3 cups warm water

Morning:

½ cup warm water

4 teaspoons active dry baking yeast (2 packets)

2 teaspoons salt

2½ to 4 cups whole-wheat flour

In the afternoon or evening: Stir the yeast and flour into the water and beat (short strokes in and out of the batter) about 100 strokes. Cover and set aside until morning. It needn't be kept at any particular temperature.

In the morning: Make sure the half cup of warm water is cool enough so that it will not harm the yeast – about body temperature. Stir in the yeast and let it dissolve, then mix this into the batter from last night. Stir in the salt as well.

Fold in about 2½ cups of whole-wheat flour a half cup at a time. Turn out onto a floured board and knead, using another half cup or more of flour to keep the dough from sticking. Knead 150 to 300 times. The dough will be smooth and pliable. Set aside in an oiled bowl and let rise 3 to 4 hours.

Shape dough into 2 loaves. I like to make log shapes with diagonal cuts on the surface, or you can bake in loaf pans if you prefer. Place loaves on oiled sheet pan and brush with water. Let rise about an hour. Brush the surface once more with water and bake in a preheated 375-degree oven 45 to 60 minutes until the loaves are browned top and bottom.

Secrets Rarely Revealed: The Pots Come Clean

WHEN I HAD the opportunity to work in the kitchen at Tassajara Hot Springs in May 1966, I didn't hesitate. Compared to the study of logic and statistics, psychology and sociology, which I was pursuing in college, the idea of a life of cooking had its appeal. I had gotten an "A" on my paper about alienation and anxiety, and I was just as alienated and anxious as ever. That spring, book learning didn't seem worth much, whereas cooking seemed so real and practical, down-to-earth and enjoyable. At Tassajara – still a resort owned by Bob and Anna Beck – I started as the dishwasher, cleaning all the pots and pans, mixing bowls, and utensils as well as all the dishes. The dishes I'd wash by hand before placing them in racks to go through a machine that gave them a sterilizing rinse. I figured out several things rather quickly to make my job easier. I felt mellow and competent. I worked by myself, did things my way, got them done, went swimming and lay in the sun. Life was beautiful.

The things I learned about washing dishes and cleaning pots were simple enough, but since then I've noticed how often people do things the hard way. I find it a mystery. In washing more than a few dishes one point is pivotal: Sort the dishes as early as possible in the process. Curiously, the dish-washing machine is the slowest part of the activity in a commercial setting; a person racking and unracking the dishes can work faster than the machine. So the secret of washing dishes is to make the machine work more efficiently by putting more dishes in the rack each time through the machine. At home the same principle applies: Sorting and stacking the dishes will greatly aid the dish-washing process, whether one is doing them by hand or trying to fit dishes into a dishwasher.

Often people's inclination is to grab the dishes as they come in and pile them onto the racks or into the sink without sorting them. Yet the dishes are going to have to be sorted sooner or later, since they go back on the shelves stacked in piles of the same size. When the dishes are sorted first, they stay sorted the rest of the way through the process; the dish rack can be filled fuller; and, once they are dry, the dishes can be stacked much faster: zip, zip, zip.

Another advantage of sorting the dishes first is that it keeps space clear for more dirty dishes to come in, which otherwise might pile up in awkward places and spill, crash, obstruct. Why make things so difficult?

At Tassajara the setup for cleaning the pots and other kitchen items included a counter where the dirty items were placed, a double sink for soapy water and rinse water, and some slatted shelves for draining things. Another sink nearby contained a large cone-shaped sieve, so that waste could be poured into it and drained, leaving solid matter to be discarded. This proved to be extremely useful.

Having worked in a great number of kitchens, I have noticed that people rarely understand what is pivotal in pot washing, but the fundamental secret is to keep the wash water clean! When the soapy wash water is permeated with minestrone soup, pan drippings, and salad dressing, it will make things dirtier (certainly greasier) than they were, and you will find the work aesthetically unpleasing and unappetizing.

I cannot think of a single good reason to do that to yourself when a simple and elegant solution is at hand, which, with marvelous coincidence, also solves a second pot-washing annoyance: When big pots and pans are washed in a big sink of soapy water, they can't really be scrubbed effectively, because everything is bobbing around in the water, and endlessly bending over the sink can give you a backache.

So, prerinse and prescrub everything on the *solid* counter next to the sink (or even on the floor if you are faced with a really giant pot) and then pour this particle-enriched water through a cone sieve, preferably set in a separate sink. Alternatives to the sieve-in-a-sink approach are to have a spritzer nozzle that can be used to spray things off or a garbage disposal. Once the pots and pans are "clean," they are ready to be washed and rinsed. This part goes really quickly; you race right through it; and you have saved yourself from washing everything in soup.

Okay, so I was a brilliant dishwasher. I worked hard, learned to bake bread as well, and felt calm and serene. When the cooks screamed or threw fits (which was comparatively rare), I'd shake my head. "What's their problem?" I'd wonder. "If I was a cook, I'd never behave like that!" About halfway through the summer one of the cooks quit, and I was offered his job. Within two or three days I too was screaming at times. The cool dishwasher was transformed into a hot-under-the-collar cook. "Cook's temperament" comes with the territory.

Equipment

A good pot- and dish-washing setup will contribute to the efficiency of a kitchen and consequently the buoyancy of the cook. While there is a great array of equipment available for kitchen work, I find a few items particularly useful. So I want to tell you about them, even though I tend to skip over the equipment sections of cookbooks. Over time you will discover which items make all the difference for you between drudgery and joyful endeavor.

KNIFE AND CUTTING BOARD

When I began cooking in the sixties, I would sometimes visit people's houses and notice with dismay a small old paring knife and look fruitlessly for a cutting board. I wondered how they could do much in the way of cooking, and usually they didn't. Once you find a knife you like to work with, learn to take care of it, and develop some facility with it, you are all the more ready and equipped to handle food.

I use a Japanese vegetable-cutting knife, which is designed differently from a Western chef's knife. The Western chef's knife is designed primarily to cut meat and only secondarily to cut vegetables – the blade has a "vee" shape, which helps to wedge the flesh apart. A Japanese vegetable-cutting knife has sides that are closer to parallel, with a longer taper, making the blade sharper and narrower. Of course, just as with chef's knives, the Japanese-style knife comes in a wide range of quality. The one I use is made from two kinds of steel, a core of high-carbon steel, which can be made exceedingly sharp, and a sleeve of more durable, less brittle steel.

Well, I could go on and on about my knife, but the important thing is for you to find a knife you like and a cutting board as well. I prefer the wooden ones. For a while, white plastic cutting boards were being touted as cleaner and more hygienic than wood, but more recent studies have shown that wood actually harbors fewer bacteria. Those made of good-quality hardwoods will not gouge or splinter, and will be easy to clean. Periodically I scrub mine with soap and water, rinse, dry, and soak with mineral oil.

ELECTRIC COFFEE MILL OR SPICE GRINDER

Using an electric mill to grind freshly your herbs and spices provides more intense, immediate flavors. The spices in those jars that have been on the shelf for longer than you can remember have no flavor left in them.

This alone can make a tremendous difference in your cooking, as you can grind cumin, coriander, or fennel seeds; cinnamon, cloves, allspice, or anise; cardamom; and dried chilies. I also use my coffee mill to grind nuts and sesame seeds into meal for use in sauces and desserts. Occasionally I even use it to grind lemon or orange peel, stripped from the fruit with a vegetable peeler, into zest.

I recommend an electric coffee mill rather than a spice grinder because the spice grinders have a less sturdy construction, which is not nearly so functional – what's the point? And if you also like to grind your own coffee, I suggest you get two coffee mills, so that the coffee flavors do not ruin the flavor of the spices and vice versa. (I prefer Krups by the way.)

HEAT-RESISTANT RUBBER SPATULAS

The ones I like to use are made by Rubbermaid, and they have become an endangered species, because cheaper, less useful spatulas have flooded the market. "Heat-resistant" means they can be used to clean out a hot pan or pot, or to stir things, like garlic and ginger for instance, off the bottom of the pan when cooking. They are firm enough to apply pressure for "scraping" and flexible enough to bend around between avocado flesh and peel, so that the flesh pops out whole and beautiful.

Spatulas that are not truly heat-resistant – and some manufacturers have the audacity to say "heat-resistant" and in smaller print: "Caution: will melt if in contact with hot metal" – are often of such flimsy construction as to be useless for scraping.

The Rubbermaid heat-resistant rubber spatula used to be in all the supermarkets, and now I am lucky to find them at all! The problem seems to be that people tend to choose a useless rubber spatula for half the price rather than one that is truly useful. Yet once people at my cooking classes realize how handy they are, they become converts.

IMMERSION BLENDER

A caterer friend gave me her old Cuisinart when she upgraded to a Super-Pro. I found it quite useful at times for blending soups and making multiple batches of tart dough. Still, I have a minimum of counter space, and mostly the Cuisinart sits on the bottom shelf out of the way.

For blending I turn to my immersion blender, which is also known as a "blender-on-a-stick." It is quite practical and effective for most uses, as you can plug it in and then immerse it in soup or sauce to blend, without having the additional dirty food processor, bowls, or pots to be cleaned. It is also easy to rinse off. Braun makes the one I use, which comes with a convenient wall-mount.

NONSTICK FRYING PANS OR SKILLETS

I resisted Teflon for a long time but now have a couple of pans I use for making crêpes. (If you use this kind of pan for crêpes, you will find the heat-resistant rubber spatula helpful for lifting the edge of the crêpe. Some of my pans are seriously scratched from cooking classes where students tired of using the spatula and switched to metal implements.)

In the last year or two I have finally invested in some heavy-duty nonstick frying pans with little circular ridges. They are fabulously useful – for frying potatoes, if nothing else, as well as for heating leftovers. With a small amount of water and a cover, leftovers can be quickly heated and then turned right out of the pan (if you use a heat-resistant rubber spatula, anyway).

PRESSURE COOKER

I find a pressure cooker very useful for cooking beans, as this method takes a third to a quarter of the time that cooking in a regular pot takes. Lately I can't find mine. I wonder if I left it at a cooking class somewhere. If I don't find it soon, I'll have to buy another.

First Aid

I take Band-Aids with me to all cooking events, as many people keep empty boxes, thinking that they have Band-Aids on hand.

I also keep a couple of Chinese herbal remedies close by. Yunnan Pai Yao is an excellent and effective Chinese herbal remedy for helping wounds heal. It helps to stop bleeding and prevent bruising and swelling. Friends and I have had great success with it, so I keep it on hand for any knife wounds. It is also a recommended aid for recovery from surgery or for fingers smashed in car doors. It's available from Chinese herbal stores and often from Chinese groceries, and it's also starting to appear in natural food stores. I found out about Yunnan Pai Yao when I worked in a Chinese medicine clinic for a year.

Perhaps you already have a favored burn ointment, but I use a Chinese herbal burn ointment called Ching Wan Hung. We use this now at Tassajara as well. In a pinch you can also use honey.

Sharpen Your Mind to Sharpen the Knife

For many summers I have given a workshop for guests at Tassajara called "Zen and Cooking" or "Cooking as a Spiritual Practice." I come down a few days early to prepare and usually spend a while in the kitchen giving residents vegetable-cutting and knife-sharpening demonstrations.

I love cutting vegetables, and part of the reason I love cutting vegetables is because I have sharp knives. When I have to work with a dull knife, I find that cutting vegetables is tiresome and tedious. Since I want to go on loving to cut vegetables, I go on sharpening my knives. Also I know to take my knives with me if I am going some place where I might be asked to cook, because most kitchens do not have sharp knives.

The kitchen at Tassajara is no exception. When knives are shared, one person is assigned to be knife sharpener. This person usually has other responsibilities, and knife sharpening, viewed as being dull or boring, is not an especially high priority. Knives remain dull because the activity of sharpening is considered to be dull. And the duller the knives become, the more work is required to resharpen them. It's easier to sharpen a sharp knife than a dull one.

The frustrated knife sharpener decides that the knives must be dull because of the way people are using them (not because of their lack of sharpening). So the knife sharpener proceeds to make rules about how the knives may and may not be used.

I know that to keep a knife in good shape, certainly I need to be cautious about how I use it. I know, for instance, how I ruined one of my knives by using it to cut up an old Styrofoam cooler that wouldn't fit into the trash can until I sawed it into smaller pieces. That's pretty obvious. But what the knife sharpeners decide is that it's against the rules to use the knife to scrape the vegetables sideways on the cutting board. "What you have to do," they say, "is to put the knife down and pick up a metal scraper for that pushing-the-vegetables-to-the-side stuff."

Personally I find using a knife for both cutting and pushing to be eminently practical. I would find it terribly annoying to have to put down my knife and pick up some other implement to push the cut pieces aside. So after I have done some cutting, I put the cutting edge of the knife flush against the cutting board and push the cut pieces to the side to make more space to cut. Or I use the knife to scrape the cut pieces into a bowl. Simple.

So when I get to Tassajara and do my vegetable cutting demonstration: slice, slice, slice . . . scrape; slice, slice, slice . . . scrape; I can sense the students' fascination with my dexterity and also their befuddlement at my flaunting of the rules. I know the question is coming, and I can hardly wait.

"Do you always do that – scrape your knife sideways on the cutting board?"

"Yes, I do."

"Doesn't that dull your knife?"

That's when I pounce. "Whose knife is sharper, yours or mine?" I ask. I have to answer my own question, because most of them haven't yet had a chance to try mine. "Mine," I tell them. "Mine is sharper because I sharpen it."

Then I elaborate: "When you sharpen your own knife, you get to use it the way you want to use it. When someone else sharpens your knife for you, he or she makes the rules." In professional kitchens each cook has his or her own knife, and one must take care of one's own sharpening.

Sharpness starts with making your mind sharp. This means approaching the activity of knife-sharpening with some keenness, which will include: focus, steadiness, investigation, energy, exactness, patience, concentration. To achieve sharpness, you steadily and energetically focus your attention on the activity, investigating how to do it, examining the results, adjusting your effort, patiently persevering, focusing. . . .

Sharpness takes time and steady awareness, or the hands will waver. Observing various indications, listening to the sound of the knife on the sharpening stone, sensing with your hands the subtle changes – this is sharpness. If a knife is really dull, working on it for a concentrated ten minutes a day for a week may be more effective than one unfocused hour. Cultivate your sharpness over time, and the results will follow. Knives will cut easily and effortlessly.

Recipes

Much of my cuisine is centered around cutting. Cut surfaces release more flavor than uncut surfaces. An obvious way to demonstrate the difference in taste is to bite into a whole apple and then bite into a slice of apple. In the former case the flavor of the apple becomes apparent only after some chewing, whereas in the latter case, the flavor immediately blossoms. So part of the reason that my food is flavorful is all the cutting I do.

Flavor is the reason so many of my instructions mention cutting on the diago-

nal: carrots, zucchini, asparagus. Each piece of a diagonally cut vegetable will have more cut surface exposed than a piece cut crosswise. Also the diagonal pieces do not stick together the way crosscut pieces do.

To cut on the diagonal, some people think they should position the knife crosswise to the vegetable and then, holding the knife at an angle of thirty or forty-five degrees, try to cut. A more effective way to cut on the diagonal is to cut straight down with the knife, but to position the vegetable so that it is almost in line with the knife.

Following is a vegetable pasta for which you can utilize or develop your cutting skills.

Multicolored Pasta and Vegetables with Dry Vermouth
Spinach Salad with Apples and Almonds

Multicolored Pasta and Vegetables with Dry Vermouth

This pasta dish is not only colorful and flavorful – the corkscrew pasta comes in a mix of egg, spinach, and beet – but also comparatively low in calories since it utilizes dry vermouth for the sauce rather than olive oil or cream.

People seem to have the capacity to eat varying amounts of pasta. In most cases a quarter pound per person makes quite a generous serving, depending, of course, on what else is on the menu.

Serves 4 to 6 people

1 pound mixed vegetables, such as carrots, green beans, mushrooms, celery, bell
      peppers of various colors, asparagus
2 tablespoons olive oil
2 small yellow onions, diced
4 cloves garlic, minced

1 cup dry vermouth

Salt and black pepper, freshly ground

1 pound "rainbow" corkscrew pasta

¼ cup fresh herbs, minced, such as parsley, thyme, marjoram, chervil, basil, or
   combination thereof

A few drops of lemon juice or balsamic vinegar (optional)

½ dozen sun-dried tomatoes, plumped and cut in thin strips

½ cup or so Parmesan or Asiago cheese, freshly grated

Start 3 to 4 quarts of water heating to blanch the vegetables and cook the pasta.

Choose a combination of vegetables that is colorful or what you enjoy or have available. Cut them into fork-sized pieces, generally narrow strips between 1 and 2 inches long. Green beans can be "French-cut" in long diagonals or cut in half lengthwise; the carrots julienned. The exception is mushrooms, which might better be left in good-sized chunks.

Heat the olive oil in a skillet large enough to hold all the ingredients. Sauté the onions a few minutes until they are translucent. Add the garlic and let it cook briefly before adding the mushrooms (if you are using them). Let them cook and release their juices, then add the vermouth, and continue cooking to reduce the liquid by half. Set aside.

Add 2 to 3 teaspoons of salt to the cooking water, and when it is boiling, blanch the vegetables individually, cooking each one until it is as tender as you like. (I do green beans about 3 minutes, carrot slivers about 1 minute, bell pepper strips about 1 minute.) Remove from the water with a slotted spoon or strainer and set aside. The cooked vegetables may be added to the large skillet, if it is off the heat.

Once the vegetables are blanched, start the pasta cooking. When the pasta is nearly cooked, start reheating all the vegetables in the skillet with the onions and vermouth. When it is tender, drain the pasta and combine it with the vegetables. Add about half the fresh herbs, then season with a few drops of lemon juice or balsamic vinegar to brighten the flavors. Little salt may be necessary considering that the sun-dried tomatoes and grated cheese are yet to come and both of these provide salt, but see what you think by tasting, and grind in some black pepper.

Garnish with the sun-dried tomatoes and the remaining herbs and serve. Pass the grated cheese separately, so that all the splendid colors may be enjoyed before they are engulfed with cheese.

Spinach Salad with Apples and Almonds

I learned about wilted spinach salads from working at Greens, where we had a fine one with feta cheese, croutons, and Kalamata olives on the menu daily. Here's one of my simplified versions.

Serves 4 to 6 people

1 bunch spinach

½ cup almonds

½ teaspoon cumin seed

½ teaspoon coriander seed

Juice of 1 lime (about ¼ cup)

¼ teaspoon salt

1 tablespoon honey

1 good eating apple, such as Gala, Fuji, Golden Delicious

1 to 2 cloves garlic, minced (optional)

¼ cup olive oil

Cut off the base of the spinach, then cut the tough stems off of the leaves. Wash the leaves thoroughly, then spin dry. Cut the biggest leaves into halves or thirds.

Roast the almonds for 7 to 8 minutes in a 350-degree oven, or pan-roast on top of the stove until crunchy when chopped. Slice the roasted almonds with a sharp knife, or chop briefly in a Cuisinart on pulse, so the almonds do not turn completely to powder.

To make the dressing, start by grinding the cumin and coriander in an electric grinder. Combine with the lime juice, salt, and honey.

Cut the apple into quarters and cut away the core. Slice the quarters into thin lengthwise pieces, and toss with the dressing.

Put the spinach in a stainless steel, wooden, or ceramic bowl. Heat the olive oil in a small pan until nearly smoking. Pour over the spinach with one hand, while using tongs to toss the spinach with the other. If the spinach is not sufficiently wilted – to your taste – press clumps of it into the hot pan using the tongs. Then toss the spinach with the apples. Check the seasoning.

Serve on a platter or individual plates and garnish with the almonds.

Liberating Your Hands

Our bodies are made to do things: to stand, walk, sit, chop wood, carry water, peel potatoes, putter about in the shop or garden. Our hands especially love to do things: knead bread, caress cheeks, dig with a shovel, fiddle with wiring, pluck guitar strings. I didn't always understand how eager my hands were to work until they met bread dough and later sharp knives.

Hands love to knit and sew, hammer and saw, knead and shape. Hands that are idle grow restless or bored and, worse, start to feel useless and unappreciated. It's not the body that's weak, but the mind. Hands love to do what hands can do, but the mind often says, "No, you can't. I don't feel like it." The hands are ready, the mind is reluctant.

I often demonstrate techniques for cutting vegetables, but watching me cut vegetables is not going to teach you how to do it. The best way is to learn as I did, by putting your awareness in your hands and actually feeling what the hands are doing. You'll notice that the hands do well when you are not telling them what to do, but letting them feel out for themselves what seems like a good way to do something and letting them try various ways and methods. Hands like that. They don't appreciate being bossed around.

As you give your attention to your hands and let them try out a variety of ways for cutting vegetables, you will notice many things. One thing I have learned over the years is to hold the knife with my thumb and first finger on either side of the blade with the other three fingers wrapped around the handle. This has at least two distinct advantages over holding the knife by the handle (often with the index finger sticking out on top of the blade): One is that the knife cannot unexpectedly twist in your hand (you are gripping it on both sides of the blade), and the other is that moving your hand forward on the knife – gripping it on the blade (like "choking up" on a baseball bat or a hammer) – gives you greater facility. Smaller movements of the hand and arm make bigger movements of the knife. Once you get used to it, you never go back.

Probably the single most useful skill to learn in cutting vegetables is to use the knife as a saw as well as to use it as a chisel, which is what most people tend to do. To use the knife as a chisel is to chop straight down, while using the knife as a saw means slicing forward and back. Using a chisel (or a chisellike approach) takes a tremendous amount of energy – a solid blow of the hammer on the end of a chisel or forceful downward pressure on the knife handle. In comparison sliding a saw (or a knife) back and forth is restful and easy. You allow the sharpness of the blade to work for you – by sliding it across the vegetable. To combine the sawing motion with the chopping down motion is most effective.

Another pivotally important point is how you hold your left (or nonknife) hand: Each and every fingertip is curled slightly back, so that no fingertips are left sticking out straight where the knife can find them! Then I use the knuckle of the middle finger of the left hand to guide the knife – the side of the knife blade rests against that knuckle and cuts where that knuckle is. People watching ask, "How can you do that? Aren't you going to cut yourself?"

The flat side of the knife is not going to cut anything, no matter how much it rubs against your knuckle, so the key to avoid cutting yourself is not to lift the cutting edge of the knife above your knuckle. Above the vegetable, OK, but not above the knuckle.

Finding Out How

Here's how it works. Start with the knife on the far side of a celery stalk, say, holding the tip of the knife on the counter with the cutting edge poised above the celery (well below the knuckle of your left hand). Stroke down and forward, or down and toward you – you've cut a slice of celery. Now what? Most people want to pick the knife *up*, completely off the table. Then, sure enough, the knife is poised above the celery *and* above your knuckles.

To avoid this problem you *leave the tip of the knife on the counter* after making the cutting stroke, and slide the knife back into position just above the celery, well below your knuckle. Cut again. As long as you don't raise the cutting edge of the knife above your knuckle, there is no way to cut yourself, and your left hand can guide the knife. If you want thin slices your left hand moves back just a little; if you want thicker pieces your left hand moves back further between strokes of the knife. Your left hand guides. Your right hand just cuts.

Hands have amazing "intelligence," so while they are cutting, the eyes do not necessarily have to be involved. About a third of the sensory neurons running from the body to the brain come from the hands, and about a third of the motor neurons spreading from the brain to the body go to the hands. That's a tremendous amount of potential intelligence, which you activate by allowing your hands to feel out what to do instead of telling them what to do.

Besides, although using your eyes to guide your hands is sometimes necessary, often it's extra, and you will just make your hands nervous. After all, how do you feel when someone watches you do something? Often we say, "Do you have to watch me like that? You're making me nervous." Similarly your hands may become awkward and clumsy when you try to guide them with your eyes. Over time you can learn to trust that your hands know really well how to be hands. When the eyes are not busy overdirecting, then they can work in harmony with the hands.

Letting your hands find out how to be hands is vitally important – it is a form of liberation. Then they can resume their natural responsiveness to things and with it their energy, joy, and vitality. Their inherent love and helpfulness emerges. When the hands are happy, that can make a being pretty happy, too.

Recipes

Here are some handy things to do with a knife – and perhaps another implement or two.

Mincing
Mincing Garlic
Making Lemon, Lime, or Orange Zest
Avocado Slices for Composition Platters

Mincing

To mince fresh herbs, pitted olives, citrus peel, or nuts, this is the basic procedure: With the ingredient on the counter in front of you, position the knife crosswise in front of you and stabilize the tip end of the blade with your left hand. You can do this by placing the fingers of your left hand out flat on top of the knife blade, or by gripping the blade near the tip between thumb and first finger. With the blade stabilized by the left hand, use the right to chop, "walking" the knife away from you between chops, and then back toward you. Slightly tilting the top of the knife in the direction of the "walk" seems to help. Stop periodically and scrape the pile back into place.

Mincing Garlic

Mincing garlic rates a special note, since it is done so often. Personally I got tired of garlic presses many years ago for two main reasons: A lot of the garlic seemed to pop up around the sides of the lever press instead of going through the little holes. Second, the garlic press was a nuisance to clean.

I find my knife quite sufficient. To peel the garlic clove I lay the side of my knife flat on top of the clove and tap it with my left fist, enough to break open the peel around the clove. Then I can easily slip the peel off.

The key for mincing garlic is getting it started. On TV I think I have seen Martin Yan of "Yan Can Cook" demonstrate pressing the flat of the knife on the garlic clove and giving it a good wallop with the left hand to smash the clove. Then it can be minced.

I like to chop the clove of garlic into thin "slices" with the *back* of my knife first, and then mince. People love this method when I demonstrate it. The keys are: hold the knife with the cutting edge directly up (upside down); position the clove of garlic close to the near edge of the cutting board; hold the garlic clove firmly in your left hand; and, then, with the back tip of the knife on the counter, lever the back of the knife down onto the clove of garlic, "cutting" and "smashing" a section off the edge. Repeat until the clove is "pre-minced," then mince it more thoroughly.

Making Lemon, Lime, or Orange Zest

1 lemon, lime, or orange

The fragrant outer layer of the rind is called zest, and whether it is minced or grated, it will provide a flavorful seasoning for soups, salad dressings, pasta dishes, desserts, and numerous other dishes. Using a grater for this seemed to leave most of the zest in the grater, so now I do it with a vegetable peeler and a sharp knife.

First of all, wash the fruit. If I am uncertain of the source of the fruit (whether it is organic or not), I use hot water to try to remove whatever may have been sprayed on its surface. If the fruit feels really waxy, I pop it in boiling water briefly.

Using a vegetable peeler, remove the colored outer layer of the peel in strips. Arrange a pile of strips on the counter and cut them crosswise into very narrow strips with quick, down-and-forward strokes of the knife. If you want the peel in even smaller pieces, mince the narrow strips.

Avocado Slices for Composition Platters

1 ripe avocado

Pick an avocado that gives slightly when gently pressed. Often you will need to buy firm avocados and allow them to ripen at home.

Cut the avocado in half lengthwise around the pit. Twist to remove half of the avocado. Whack your knife into the pit and twist to remove it. Pull the pit off the knife, or, leveraging the pit at the edge of a counter, pull the knife off the pit.

I use a heat-resistant Rubbermaid spatula to remove the avocado flesh from the skin because it is firm yet flexible enough to conform to the shape of the avocado as you insert it between the flesh and peel. Carefully work the spatula around under the avocado flesh to remove the half avocado intact. If you do not have a suitable rubber spatula for this, you may be able to use a large spoon to good effect, if you work it around slowly so as not to cut into the flesh.

To cut the avocado halves, place them flat side down on the counter, and make diagonal slices perhaps ⅛ inch thick. To do this, position the avocado at an angle to the knife and cut straight down. After each cut, hold the cut piece in place with your left hand, while you lift the knife with your right. This will keep all the cut pieces in place.

When you have finished cutting the avocado half, pick it up and place it on a plate or platter and fan out the slices.

Are You Worth Fruit for Breakfast?

A cook's sensibility extends beyond food to people and space, so in addition to providing food, a cook might provide others the opportunity to nourish themselves.

At a meditation retreat, I was passing through the buffet line for breakfast, and at the end of the serving table I came to a large bowl of fresh fruit. This is rather common at group gatherings, because it is a convenient way to offer fruit. Nothing except washing the fruit and putting it in a bowl is required. Then people have the opportunity to choose just exactly which fruit they desire.

Peels were piled on the table next to the fruit bowl, and although I could see an occasional apple core or pear stem and a few orange and tangerine peels, mostly what I noticed was a vast mound of banana peels. Most breakfasters were banana lovers. Either that, or they found that peeling a banana was about all the effort they cared to make.

"An orange would be nice, but then I'd have to peel it." "An apple would be great, but then I'd have to bite and chew it – it would be better sliced." Lots of bananas get eaten this way. We get to find out whether or not we are worth the effort it takes to open, peel, and cut a piece of fruit.

Also, the fruit setup itself was not particularly encouraging. Alongside the large bowl of fruit, the people setting up the dining room had placed a rather small plastic cutting board and two small paring knives, hardly adequate for the sixty to eighty people passing through the line. Also, since no provision had been made for parings, they piled up on the table next to the cutting board and overflowed onto the cutting board itself – really appetizing! This self-help approach to cooking brings the work of the kitchen and the mess of the kitchen to the dining room. Yet even at this rudimentary self-help level of cuisine, a cook thinks about ways to encourage people to help themselves, such as putting out a much larger cutting board at which two or more people might be able to work. An adequately sized bowl for parings (or compost buckets under the table) would also be appropriate, since the fruit has peels and skins, seeds and pits, stems and bruises, which need to be discarded.

When we stop to consider the matter, we understand that the fruit is not just there for looks, but for eating. Eating means work, and work with food means trash. We know that, so we'll make a place for people to work and a place for the trash to keep the table and work space free of trash. You are *invited* to work, given a place to do it, and the bowl for trash lets you know that you can make a mess (we expect it) and that the work space itself should be left clear, available to the next person for work.

To consider how others will experience the situation expands the mind of a cook. The bountiful opulence of cut fruit, glistening brightly, is much more appealing than whole fruit. At the same time, work requires space and time, and tools. Make it possible and inviting, and all of us are much more likely to do it.

Sometimes we are not sure: "Am I really worth sliced fruit today?" A friend says her mother used to tease her when she did something for herself: "My, aren't we feeling good about ourselves today?" Well I say, "What a fine thing to do! – for yourself or for loved ones."

Yes, we are feeling good about ourselves today, and we are worth fruit for breakfast.

Recipes

Speaking of fruit for breakfast, I love sliced oranges or sometimes – when I am really feeling good about myself – grapefruit "supremes." This does not involve genius creativity, but a sharp knife helps. Supremes are the wedges of orange or grapefruit without any of their "packing material": the peel, pith, and membrane. And no, contrary to what I believed for many years, this is not done by dividing the orange or grapefruit into segments, and then peeling the membranes off each one individually. Although for Valentine's Day, a special holiday, or a birthday breakfast, orange or grapefruit supremes are a real demonstration of devotion.

These recipes are fairly basic, but that's the point. Breakfast needn't be fancy, and sliced fruit has a reassuring appeal.

Orange Slices or Supremes
Warm Pear Slices with Tangerines
Plumped Apple Slices with Cinnamon

Orange Slices or Supremes: Plain or Elaborated

This is the more or less classical method, which I use for peeling oranges or grape-fruits for fresh fruit or salads.

1 orange per person

To peel with a knife, so that the outside of the orange (or grapefruit) has no white, cut off the top and bottom of the orange with a sharp knife, so that a circle of orange shows. Place the orange flat side down, and then cut off a section of peel, top to bottom, following the outline of the orange. Once you have cut off one section of peel, you can cut off the remaining sections of peel following the line where the orange meets the white. After you have cut off the peel all the way around, turn the orange over and trim off any parts of the peel you may have missed.

Slice the orange crosswise into rounds, or cut the orange in half vertically first and then crosswise into half-rounds.

The oranges may be served as is, or combined with slices of apple, banana, and/or raisins or date pieces. In the summer, sliced oranges are also wonderful with blueberries and strawberries or other summer fruits. I add a touch of sugar or a couple of spoonfuls of maple syrup.

With a sharp knife and a little more work you can make "supremes." Start by cutting the peel off the orange or grapefruit as in the directions above. Then hold the orange or grapefruit over a bowl to catch the juice, and cut toward the center along one side of a segment next to the membrane and then along the other side of the segment, so that the segment comes loose from the membrane on both sides. Turn the fruit in your hand and cut on each side of the next segment so that it drops loose. Continue until all the segments have been cut loose.

These may be served as is or sweetened slightly. The orange and grapefruit may be combined, and other fruits may be added.

Supremes are also an excellent way to prepare orange or grapefruit for salads.

Warm Pear Slices with Tangerines

I find this combination enjoyable in the autumn or over the holidays, when pears and tangerines are in season.

Serves 4 people

4 pears
2 tablespoons butter
1 tablespoon sugar
½ teaspoon powdered cloves
4 tangerines, peeled and segmented

Quarter, core, and slice the pears. (I don't usually peel them unless I am uncertain about their derivation.) Heat in saucepot with the butter, sugar, and powdered cloves for 4 to 5 minutes. Remove from heat and combine with tangerines.

Plumped Apple Slices with Cinnamon

This dish has become a great favorite in my household, perhaps because it is reminiscent of apple pie or apple crisp without all the butterfat – simple, yet simply delicious.

Serves 4 people

4 apples, quartered, cored, and sliced

1 tablespoon butter

1 tablespoon sugar

1 teaspoon cinnamon

Combine apple slices in a covered pot with the other ingredients and heat over medium-low heat. Give it about 10 minutes. Then stir so that ingredients are well combined.

Seeing Virtue

To See Virtue, You Have to Have a Calm Mind

When I began working as a cook in 1966, I developed cook's temperament within a few hours. As the dishwasher I had been calm and serene, and when the cooks threw an occasional tantrum, I was amused and a bit embarrassed. I had trouble believing that cooks could actually scream as venomously as they did, when it was obviously so ineffective in getting results. "That's stupid and ridiculous," I would gently remark to myself, while perhaps lifting an eyebrow. Well, I ate my words.

Sometimes the people working with you are too polite to actually confront you about your behavior, but you know for sure that others have noticed when they start having meetings to discuss "What are we going to do about Ed?" Two people were needed to replace me as dishwasher/baker, but that didn't help me to relax in my new position. "Get these eggs out while they are hot!" I would bellow. After all, shouldn't everything be perfect? And shouldn't everyone be doing their utmost and more to make it that way? At the urging of my fellow workers I agreed to make efforts to calm down.

In December of that year Tassajara was bought by Zen Center. Because I was already a zen student and had more than two months of experience cooking, I was offered the position of head cook for the new center. I made it up as I went along, and everybody knew that the kitchen procedures were not very well worked out. But I took refuge in just doing what I was doing: "When you wash the rice, wash the rice; when you stir the soup, stir the soup. . . ."

I realized pretty early on what every cook realizes: The food more or less takes care of itself; the people are what's hard. They don't do what you want. They don't behave the way you would like them to. They don't treat you the way you want to be treated. They point out your faults . . . over and over again. They won't put up with you and the repertoire of coping behaviors you've worked out. They don't applaud your every move. (For goodness sake, they aren't Mom and Dad. They want *you* to be *their* Mom and Dad.) They don't read your mind. Good grief, you have to talk with them.

The women with whom I worked were especially likely to object to my style of management:

"Why are you talking to me like that?"

"Like what?"

"Like you were angry with me about something. What have I done?"

"Look, I'm under a lot of pressure, OK? Can we just concentrate on getting the work done and not analyze every word?"

People sometimes came late to work, took long breaks, and often when I watched them working, didn't seem to be very present in their activity. I couldn't tell what they were doing, but the rice would take a long time to get washed. Finally, one day I complained to Suzuki Roshi. I told him all the problems I had with people not behaving the way I thought they ought to behave (if they were really practicing Zen): arriving late, taking long bathroom breaks, gossiping, being absentminded or inattentive. Then I asked him for advice on how to get everyone to work with more concentration and vigor.

He seemed to listen quite carefully, as though he understood my difficulty and was entirely sympathetic. (Yes, you just can't get good help anymore, can you?) When I finally ran out of complaints he looked at me briefly, and then responded, "If you want to see virtue," he said, "you have to have a calm mind."

"That isn't what I asked you," I thought to myself, but I kept quiet. I gave it some time to turn me around. Was I going to spend my time finding fault or seeing virtue? It had never occurred to me that I could spend my time seeing virtue, but my teacher's mentioning it made it seem obvious.

Later in our conversation he said, "When you are cooking, you're not just working on food. You're working on yourself. You're working on other people." Well, of course, I thought, that makes sense.

Without really having any idea how to actually do it, I began to try "to see virtue." Whenever I found fault with someone, I would remind myself to look again, more carefully and more calmly. I began to recognize people's basic good intention, to sense people's effort, the effort it took even to stand on-the-spot and be exposed for all the world to see. I would catch glimpses of our shared vulnerability.

It got to be quite laughable at times. Once I asked someone to get 18 cups of black beans from the storeroom. About twenty minutes later I realized he hadn't come back. "How difficult can it be to get eighteen cups of beans?" I righteously raged to myself as I headed for the storeroom. Yet before arriving I cautioned myself to look for virtue: What was going on? Sure enough, there he was, sorting through the beans, pretty much one by one, making sure that each was not a stone.

I felt a surge of impatience, and then I thought, "Well, he is being thorough! He is being conscientious!" I don't remember what I said, but my response was at least somewhat softened from what it would have been. Something more articulate than, "You idiot!!" emerged from my lips, and then I explained that he could cover a white plate with beans and easily scan through to check for small stones. Perhaps the sorting would go a bit more quickly that way.

Ironically, seeing virtue cultivates virtue. If we want to bring out the best in others, it helps to see the best in them. After a while we might even acknowledge the best in ourselves. A lot of struggles were still ahead of me, but over the years I have continued to cultivate my capacity to see virtue. While it's an ongoing challenge, by seeing virtue we can transform ourselves and the world.

Recipes

One aspect of seeing virtue is appreciating the simple goodness of fruits and vegetables, black beans, and winter greens. Another is respecting honest, thorough, and careful effort. The recipes here are more elaborate than most of the others in this book, yet that complexity may be seen as an expression of seeing virtue. When you have the time and take the "trouble," common ingredients well prepared can be quite delicious.

Chili Crêpes with Goat Cheese Filling
   served over
Black Beans with Garlic and Cumin
Winter Green Salad with Walnuts
   and
Roasted Red Pepper Sauce
Roasted Red Peppers

Chili Crêpes with Goat Cheese Filling served over Black Beans with Garlic and Cumin

This dish is unusual in that the crêpes are made primarily with ground chili rather than flour. Also of interest is the fact that the four main ingredients of commercial chili powders – chili, garlic, cumin, and oregano – are here divided among the different elements of the dish: The chili is in the crêpes; garlic and oregano are in the goat cheese filling; and garlic and cumin are in the black beans.

We will begin by cooking the black beans since they take the longest.

Black Beans with Garlic and Cumin

Serves 4 people generously

1 cup black beans

5 cups water

1 tablespoon whole cumin seed

1 tablespoon olive oil

1 yellow onion, diced

4 cloves garlic, minced

Salt

Sort through the beans, removing any stones, and then put the beans in a pot with the water. Let them soak overnight or for several hours during the day. Often people suggest changing the water after soaking (they say the beans cause less gas that way) but I cook my beans in the same water. Cover and heat to boiling, then reduce the heat to low, so that the water continues to bubble slightly. I peek inside now and again to make sure the water is bubbling and that there still *is* water. The beans will cook in about 45 minutes, or you can pressure-cook them in 25 to 30 minutes. Since I usually do not remember to soak the beans ahead of time, I just cook them endlessly, about 2 hours. They should be soft through and through.

Grind the whole cumin seeds in a spice mill or extra coffee grinder. Heat the oil in a skillet and sauté the onion for several minutes before adding the garlic and ground cumin. Cook another minute or so. Once the beans are cooked, add the onions to them. (Put a few spoonfuls of water in the bottom of the skillet to rinse out those good flavors, and add to the beans.)

Season with salt.

Chili Crêpes with Goat Cheese Filling

While the beans are cooking, make the crêpes and the filling.

Serves 4 people generously

The crêpes:

2 eggs

1⅓ cup milk

3 to 4 tablespoons unbleached white flour

⅓ cup ground ancho chilies (see page 19) *or* other chilies

Pinch of salt

Butter (optional)

The filling:

10 ounces goat cheese (chèvre)

¼ cup milk

1 tablespoon dried oregano

2 teaspoons lemon peel, grated or finely minced

2 cloves garlic, minced

2 shallots, finely diced

Salt

Cilantro leaves, for garnish

To make the crêpes, whisk together the eggs, milk, 3 tablespoons of flour, the ground chili, and the salt. If the mixture seems too thin at some point, add the extra tablespoon of flour. This will partly depend on the size of the eggs you are using.

Heat a 6- or 8-inch crêpe pan (or Silverstone or Teflon-coated pan) over medium-high heat until drops of water sizzle and jump about. For the first crêpe I use a bit of butter to coat the bottom of the pan, even if I am using a nonstick pan. Use about 3 tablespoons of batter per crêpe – you may need a bit more for the larger pan.

Holding the pan in one hand, pour in the measured amount of batter, and immediately tilt the pan this way and that, so that the batter coats the bottom of the pan. Cook until the top is dry and full of air holes (a minute or two), then turn and cook briefly on the second side. Remove to a plate. (The crêpe instructions on page 261 are somewhat more complete.)

Continue cooking crêpes until you have used all the batter. After the first one, the pan is "seasoned," and you can usually make several more without any additional butter. One or 2 crêpes may not turn out very well while you are getting the heat adjusted and becoming familiar with the procedure. Keep trying. The crêpes may be piled up on the plate. Hopefully, you have ended up with 6 or 8 usable crêpes, depending on their size.

To make the filling, combine the goat cheese with the milk so that it is softened, and then mix in the oregano, lemon, garlic, and shallots. If shallots are not available, you could thinly slice and sauté 3 or 4 green onions. Season to taste with salt.

To assemble the crêpes, place each crêpe with its more beautiful side down on a counter, arrange a "log" of goat cheese filling across the middle of it, then wrap the crêpe around the filling. If necessary, portion out the goat cheese so that you have enough for all the crêpes.

Place the crêpes on a baking sheet, and brush them with a little milk or water so they will not dry out. When ready, heat for 10 to 15 minutes in a preheated 375-degree oven.

To serve, ladle some black beans onto individual plates, place a crêpe or two on top, and garnish with cilantro leaves.

Winter Green Salad
with Walnuts and Roasted Red Pepper Sauce

This is an attractive, colorful dish I usually make over the holiday season since it features greens and reds. The roasted red pepper sauce is pooled around the winter greens, which have been lightly cooked to soften and sweeten them.

Serves 4 to 6 people

Roasted Red Pepper Sauce (see recipe page 81)

1 head frisse *or* ½ head curly endive *or* chickory

2 cloves garlic, minced

3 to 4 tablespoons olive oil

1 tablespoon balsamic vinegar

Salt and freshly ground black pepper

¼ cup walnuts

⅓ cup dried cranberries (optional)

Prepare the Roasted Red Pepper Sauce, according to the recipe on page 81.

Cut the greens first and then wash them. If using frisse, cut the larger leaves into 2 or 3 pieces. If using curly endive or chickory, discard any bruised outer leaves, cut the head lengthwise into quarters, then crosswise into 2-inch sections. Place the greens in a large metal bowl, and toss with the garlic.

Roast the walnuts in a 350-degree oven for about 8 minutes and then chop them coarsely.

Heat the olive oil in a large skillet until it just begins to smoke, then pour it over the frisse while tossing with tongs. Place clumps of greens back into the hot skillet to cook lightly, so they just soften but do not lose their shape entirely. Season with the balsamic vinegar, salt, and pepper.

If you are using the dried cranberries, plump them for a couple of minutes in very hot water, then drain.

To assemble the salad, place portions of the greens on individual plates and surround the greens with roasted red pepper sauce. Garnish with the walnuts and the cranberries, if using them.

Roasted Red Pepper Sauce

The flavor of the roasted red peppers is so marvelous just as it is that I tend to season the sauce rather mildly, but suit yourself. You can add more of the seasonings if you wish.

> *Serves 4 to 6 people*
>
> *Makes 1½ to 2 cups sauce*

3 red bell peppers

1 medium clove garlic, minced

2 to 3 pinches salt

2 to 3 teaspoons balsamic vinegar

2 to 3 teaspoons olive oil (optional)

Freshly ground black pepper

Roast the peppers over a gas flame until they are completely blackened. Then cool in a brown paper bag or covered container, and remove the core, seeds, and blackened skin. (If necessary, see the more complete instructions which follow.) Cut into a few pieces and puree in a Cuisinart or blender. Take this opportunity to taste the remarkable flavor of roasted red pepper before adding the other ingredients.

Season with the garlic, salt, and vinegar. To really know for yourself what flavor each one contributes, add one at a time and taste after each one. You may want to add some olive oil to round out the texture and flavor. Finish the seasoning with a few grindings of black pepper.

Roasted Red Peppers

Roast the red pepper(s) over a gas flame until they are completely blackened, turning as necessary. If you overdo it, the peppers will turn from black to gray, so don't worry, here's your chance to burn some food on purpose – get them good and black!

The peppers can also be blackened over a charcoal grill or in the broiler. To blacken in the broiler, place on a baking sheet under the broiler and leave until the tops of the peppers are blackened, perhaps 5 to 10 minutes, depending on the broiler – sometimes in electric-oven broilers this can take 15 to 20 minutes. Turn over and broil the other side until blackened.

Place the blackened peppers in a brown paper bag or a covered bowl or pot to steam in their own heat. When cool enough, remove core, seeds, and blackened skin – a messy job, but the results are well worth it. I like to do this by first cutting out the core and removing all the seeds. Then I wipe off the counter and lay the pepper down with its black side up and scrape off the black with a knife. Sometimes it is easy enough to remove the blackened skin by hand. Do not rinse off the blackened skin with water as this will wash away flavor as well. Keep plenty of damp sponges or towels handy for wiping the counter, the knife, and your hands.

You now have some red pepper "fillets," which you can cut into strips or squares, or blend.

Radishes Smile, and All Beings Rejoice

What makes food *food* interests me deeply. Not everything that is edible nourishes the spirit, or soothes our deeper hungers, and for food to be food it must feed. We all know this, and that each of us is nourished by particular foods. Still we forget at times that it is not just the food in front of us, but our readiness to receive, to be blessed with food, that allows food to do its feeding.

Some hungers, of course, do not respond to food, yet to eat well – and this needn't be elaborate – means the world cares. Or perhaps more precisely it means that one of the world's myriad creatures – the one who is doing the eating – feels cared for. Which brings us to radishes.

My friends Pamela and Jerry, who have a cooking school in their home, had wanted me to meet their friend Robert. At the time Robert was still involved with his restaurant in San Francisco called Le Trou. Even my daughter, who lived in France for nine years, didn't quite get the name when she came to visit.

"The Hole?" she asked, somewhat baffled.

"Yes, it's short for The Hole-in-the-Wall Restaurant," Robert explained dryly.

We had picked a date when we could all go to Le Trou, but it turned out to be a night when the restaurant was closed. No problem, we were told. Robert wished to invite us to his house for dinner. Splendid! A cook who will cook even on his day off. I felt honored already. One chef meeting another – what could be in store?

I decided to be as gracious as my wine cellar would allow, and so I brought along two bottles of ten-year-old California wines. Not knowing the menu I didn't know whether they would be appropriate for the meal, but I knew well enough I had better not count on drinking them that night. Let the chef decide.

After a bit of back and forth on a dimly lit street with tiny numbers on the houses, Patti and I found "300" and rang the doorbell. We were greeted enthusiastically by Robert and escorted up the stairs into warmth and light. Giving Robert the wine, I saw two bottles of twenty-year-old Bordeaux awaited us on the mantelpiece. "Oh fine, fine," I exclaimed, "have these another time." I was thrilled. Robert does have his ways of taking you *in*, making you comfortable and happy.

Already won over, I was still not prepared for the magnificent simplicity of our appetizers. Radishes! Seated at a low table, we came face to face with platters of radishes, brilliantly red and curvaceous, some elongated and white tipped, rootlets intact with topknots of green leaves sprouting from the opposite end. It was love at first sight. Gazing at the plentitude of radishes, red and round with narrow roots and spreading stems, I felt a swelling joy.

Perhaps I am unusual, and in fact even for me it is rare to be so awed by radishes. So often one's first response might be more like "Radishes?" or "Radishes, that's it?" or "Radishes, geez. . . ." and then perhaps a reminder to oneself, "OK, try to be polite." Yet these radishes kept growing on me, as though they exuded happiness.

Next to the platters of radishes were small dishes of sweet butter and salt. I tried a radish, first with salt, then with butter, then with butter and salt, then some plain bites – amazing, four dishes in one. Fundamental goodness, so often elusive, was plain to taste.

Accompanying the radishes was French sparkling cider, the kind that is mildly alcoholic, slightly bitter, and not especially sweet – quite refreshing with the pungent and crunchy, the creamy, the salty. The radishes also sparkled, somehow shining with the gemlike quality inherent in objects which have been removed from dirt and polished. Life reveals its preciousness.

The rest of the meal was excellent: succulent slices of roast lamb, a salad with a curry vinaigrette. The wines were indeed superb, but what remains most vividly in my heart are the radishes.

To be able to see the virtue, to appreciate the goodness of simple unadorned ingredients – this is probably the primary task of a cook. When radishes aren't good enough, pretty soon nothing is good enough. Everything falls short. Nothing measures up.

Yet when someone can pick up a radish and be delighted, this is the basis for innumerable dishes. Delight moves through radishes and people alike, letting things speak, perhaps even sing for themselves, bringing out the best at each step. A radish appears, radiantly expressing the radishness of radishes, and all beings benefit.

Recipes

The Virtue of Radishes
Radish and Carrot Salad with Green Bell Pepper
Radish Salad with Sprouts and Oranges

Growing up we mostly ate radishes off raw vegetable platters, dipping them bite by bite in small mounds of salt. My father also loved to eat raw green onions this way, and his enthusiasm for this food remains vivid in my memory.

Although there are several kinds of radishes, including daikon and black radish, the recipes here utilize the common red radish of my childhood. Quite versatile, radishes combine with other ingredients to make mild relishes for a side dish or accompaniment.

I appreciate the invigorating flavor and refreshing crispness of radishes, and guess what? In Chinese medicine they are considered to aid in digestion, helping to disperse Qi, which means that besides their culinary attributes, they will help to eliminate stagnation in digestion.

The Virtue of Radishes

The simplicity of this recipe is deceptive. Radishes in the supermarket often don't look too happy, and this dish depends on the goodness of the radishes, which probably has more to do with their upbringing than the creativity of the cook.

- Radishes, round and red, white and elongated, or red with white
  Sweet butter
  Salt

Find radishes that delight you. You might have to try a farmer's market or plant some in your yard or a windowbox.

Wash the radishes and remove the largest leaves. Arrange on a platter and serve with butter and salt in little dishes on the side.

This might be accompanied with sparkling cider or the mildly alcoholic French sparkling cider.

Radish and Carrot Salad with Green Bell Pepper

Serves 2 to 4 people

Small green bell pepper (about 4 ounces)

10 to 12 radishes (about 4 ounces)

1 large or 2 small carrots (about 4 ounces), grated

2 teaspoons honey

2 tablespoons cider or white wine vinegar

Pinch of salt

2 teaspoons dill weed

Cut the pepper in halves and remove stem and seeds. Slice each half lengthwise into 4 or 5 pieces, then crosswise into thin pieces. Slice radishes into rounds. Combine peppers and radishes with carrots.

Stir honey into vinegar to dissolve. Toss dressing with salad along with pinch of salt and dill weed. You're all set.

Radish Salad with Sprouts and Oranges

Even though I like the flavor and texture of alfalfa sprouts, I don't get them very often, but when I do, this seems to be the main way I enjoy them – soaked in orange juice. I mention that because not everyone appreciates sprouts, which are seen as being too much like "rabbit food" – hence, a sproutless variation also follows.

Serves 3 to 4 people

3 oranges

1 bunch of radishes

3 to 4 ounces alfalfa sprouts

Orange marmalade with balsamic vinegar (optional)

Cut the peel off the oranges, and slice them into rounds or half-rounds (page 67). Slice the radishes into rounds. Break up the clump of sprouts into smaller pieces and toss with the oranges and radishes. I like it like that! – a clean, refreshing flavor with no oil, salt, or pepper.

I usually get navel oranges, which are sweeter than Valencias, so if it turns out that you find the salad too tart, try sweetening it with a tablespoon or two of orange marmalade and perhaps a touch of balsamic vinegar. Drain off some of the juice, mix it with the marmalade, and re-toss.

VARIATION

In place of the alfalfa sprouts, use half a bunch of chives cut into narrow sections, or 2 or 3 green onions cut into thin pieces. To fill out the bulk of the salad, you may wish to add another orange.

Face-to-Face Encounter

Cooking is often a struggle. Anyone who has done it, whether professionally or for a family, knows. We always find more last-minute things to do than we anticipated. Everything is happening at once and needs attention. When I was cooking on a schedule, almost every meal prep went down to the wire: Would we make it? If I had extra time, I would start dreaming up more inventive things to do, including more elaborate garnishes to enhance the dishes. All the available time would be used.

Some days were more stormy than others. In those early years at Tassajara I would often work 10 to 12 hours a day, as well as attend morning and evening meditation. For a while we even did "kitchen zazen" – an extra period of meditation in midmorning just for the kitchen staff who had missed the second period of morning meditation in order to prepare breakfast. As the head cook I would work ten, twelve, fourteen days in a row. At least once I worked a month straight through. I didn't know any better.

My mental state was quite volatile at times, and I had little equanimity. I had "gone under" long before, without even realizing I was under. Being overwhelmed had become normal.

When one describes mind as space, then thoughts, feelings, emotions come as clouds, wind, rain, thunder. I didn't experience many sunny days, but one day Suzuki Roshi came to the kitchen and cleared things up. I was deeply involved in a task at the main work table, struggling to do what I was doing, while also striving to keep track of everything that needed to be done, as well as wrestling with that voice that persists in taunting, "You're never going to make it."

In the midst of this raging torrent I became aware of another voice quietly calling my name, "Ed?" At first I thought I might be hearing things. I don't know how long that voice was saying "Ed?" before I finally realized that Suzuki Roshi was standing in the doorway, calling to me. Although he was calling my name, I wasn't sure that he meant me, because his tone of voice was calling out to the kindest, most compassionate person you could ever hope to meet, while I was stormy, dark, and intense.

Who could he be calling? Several moments of befuddlement followed before I realized he was calling *me*. That good-hearted, spacious-minded person was me, and suddenly a sweet radiance permeated my body. The storm clouds vanished. The air became clear and sparkling as it does after a rain. I knew I was also this person I had never met before as well as the person struggling. He asked me about some mundane matter. Although I was attentive, I was also stupefied.

We never know how or when we will meet ourselves. We are always on the lookout for someone or something to introduce us to our more inward being, our original, bright, not-mixed-up nature. Though that clear nature can't be sustained – we are sure to have recurrent difficulties – once we know it, then the ongoing daily drama loses some of its luster as the basis for establishing selfworth. We realize we don't have to identify with our problems as the limit of who we are.

For the most part I continued being a person possessed, someone who wanted to be known as the best cook, the greatest zen student. However, no one seemed to be paying much attention, and now it didn't seem quite so important. In some fundamental way I knew I was OK.

The things that happen to us aren't the end of the world. We don't have to identify with each success or failure in the kitchen (or out) as the basis of our self-esteem. "Big mind," the Roshi would say, "is always with you, always on your side."

Recipes

Just as people can surprise us with their goodness or depth of being, so too can certain foods. Baked vegetables may sound like a prosaic dish, but they end up being surprisingly satisfying. Not every dish needs to be glamorous or special; this one is simple, yet brings our attention to the sweet wholesomeness of what appears to be nondescript. Sometimes when I cannot think of what else to make with a particular menu, baked vegetables are what fit the bill.

If I am looking for a fulfilling quick snack or light meal I often make cheese quesadillas.

Baked Vegetable Platter
Cheese Quesadillas

Baked Vegetable Platter

The plan is simple: Cut the vegetables into large pieces, toss them with a tiny bit of olive oil, and bake them. The list of vegetables is rather basic, so you could substitute with what you have on hand or what is in season.

Serves 4 people

1 yellow onion

2 carrots

2 to 3 red potatoes

1 red pepper

2 stalks celery

Several cloves garlic (optional)

1 to 2 tablespoons olive oil

A few pinches of salt

2 to 3 teaspoons or more balsamic vinegar

Fresh parsley or thyme, chopped, for garnish

For baking it's a good idea to cut the pieces into fairly large-size chunks, since thin slices will dry up in baking.

Peel the onion and cut each half into 3 or 4 wedges. Wash the carrots and cut them into 2-inch sections, and if some sections are quite fat, cut them in half. Cut each potato in halves or thirds, and each of these into quarters. Cut open the pepper, remove the core and seeds, and cut each half into 5 or 6 pieces. Bend back the wide end of the celery stalks, so that when the end breaks off you can use it to pull off some of the strings. Then cut the celery stalks into 2-inch pieces. If desired, add 2 or 3 garlic cloves per person, leaving the peel on. Place all the vegetables in a bowl and toss with the olive oil and a few pinches of salt.

Place vegetables on an oiled baking sheet, and bake in a 375-degree oven for about 45 minutes. Basically you can just leave them alone, but it's also enjoyable to check them now and again, especially since they just might need tossing and turning (depending on how your oven is treating them).

When tender, remove to a casserole or serving platter. Toss with the balsamic vinegar, check the seasoning – black pepper could also be good – and garnish with the fresh herbs.

Cheese Quesadillas

I used to make grilled cheese sandwiches, but now I'm more likely to prepare quesadillas.

Makes 1 quesadilla

1 corn tortilla
2 ounces cheese (Jack, provolone, Muenster, Gouda, or whatever you have)
1 tablespoon fresh cilantro, minced or chopped

Optional:
1 to 2 tablespoons salsa
4 to 5 slices avocado

Heat the tortilla in a dry skillet, slice the cheese, and arrange it on half of the tortilla, along with the cilantro. Fold over the other half of the tortilla and cook over moderate heat. After a half-minute or so, turn over the tortilla and cook until the cheese is melted.

Open the quesadilla and put in some salsa and avocado slices, if you like.

Nurturing the Heart

One of the primary ways we connect with each other is by eating together. Some of the connection happens simply by being in the same place at the same time and sharing the same food, but we also connect through specific actions, such as serving food to one another or making toasts: "May I offer you some potatoes?" "Here's to your health and happiness." Much of our fundamental well-being comes from the basic reassurance that there is a place for us at the table. We belong here. Here we are served and we serve others. Here we give and receive sustenance. No small matter.

I found that serving food in the meditation hall at Tassajara was an extremely powerful practice, powerful because it was a deeply intimate activity.

Taking place in silence, the basic transaction of serving food is brought to life, so that the subtle inner workings become apparent. The mind of the server and the mind of the recipient are transparently revealed – you don't have to be a genius.

Suzuki Roshi often said that when we all sat in the same posture, as we did in meditation, it was easy to tell the differences between people. Sure enough, serving one person after another, the flavor of each was apparent: anxiety, greed, calm, respect, anger, fatigue. We were all so nakedly revealed for what we were. And people receiving their food could tell the mind of the server: ease or awkwardness, nervous or composed.

Suzuki Roshi's mind was unique, vast and spacious rather than small and petty. He seemed to be neither conniving to produce particular results nor struggling to avoid other outcomes. His movements were ordinary and unremarkable, yet he was vitally present and precisely responsive. Without rushing or being hasty, his bowl would be in exactly the right spot to receive the food, to receive me. Over and over again, when I served him he was like this. A wave of tenderness would come over me: He was just there, ready to be with what came.

Once in the question-and-answer ceremony after sesshin someone asked Suzuki Roshi what he felt when she was serving him food. Yes, I thought, what is his mind at that time? "I feel like you are offering me your most complete love, your entire being," he answered, and I knew it was true, because that's what I was doing when I served him, and I knew he was receiving me thoroughly and wholeheartedly, without reservation. I felt healed each time I served him.

It wouldn't last long though. As I proceeded down the row after serving the Roshi, my more ordinary mind would return, and I'd become progressively speedier, running a silent critical commentary: "Can't you get your bowl out here more quickly? Where's your mind anyway?" "Do you have to be so greedy?" "Stop being so picky." I had something to criticize about everyone except Suzuki Roshi.

A part of our training was learning to move energetically, the Japanese Zen ideal of movement with vigor and enthusiasm. So I would try to serve as many people as I could as quickly as possible, which is not the same, you might note (as I was studiously not noting), as being polite or gracious. Basically I would be racing the server on the other side of the meditation hall to see who could finish first.

The people being served tended to get in my way and not cooperate as effectively as they might to see that I got down the row quickly. Once in a while I would remind myself to try to see virtue. "Calm down," I'd tell myself, "don't be in such a hurry to get to the end of the row." Yet this was difficult because I prided myself on being the fastest server.

I wasn't happy being caught up in this obsession, but I didn't know what to do. Then one day I had a sudden inspiration, "Why don't I treat each person as though he is Suzuki Roshi?" Is there fundamentally a real difference between people or are there just these differences that I make up and believe are important? Isn't everyone basically worthy of respect and careful attention? Why don't I treat everyone as if she is Suzuki Roshi, because each person is at some level Suzuki Roshi. I saw that I could bring the same mind I brought to serving the Roshi to serving each person: the same respect, courtesy, tenderness, and patience.

Doing this was difficult at first. By the second or third person after the Roshi my habitual mind was back in play, but gradually I slowed down. I don't know if anyone noticed a difference – no one commented to me about the change – but I felt lighter and more connected, not only to others but also to my own being. The fact that I was no longer belittling and demeaning others meant that some part of me could relax and be at ease as well, no longer in fear of being attacked. To honor the person being served is to grow larger-hearted and honor oneself as well.

I have kept up this practice for many years now, so that even when I became a waiter at Greens I continued to make this effort to serve each person as I would serve Suzuki Roshi. I did my best not to get involved with who was who, how they behaved, or how they "deserved" to be treated based on how they treated me: "Here is your food, my heartfelt offering for your well-being. May your heart be at peace, and may you grow in wisdom and compassion."

Recipes

When I think of serving Suzuki Roshi I think of what is straightforward and reassuring: white rice, tofu, and vegetables. Here's an example of how I often eat at home. Fortunately I enjoy the plain flavor of white rice. That and some fresh vegetables, and I am happy. Also, when I eat more lightly in the evening, I sleep better and wake up truly refreshed.

White Rice
Tofu with Mushrooms, Carrot, and Spinach

White Rice

Japanese white rice comes coated with talc or extra rice starch, so it needs to be washed several times until the water is clear. I usually get short-grain white Basmati rice, and make extra to have around for fried rice (see page 109), or to reheat for later.

Serves 2 to 4 people, depending on appetite and menu

1 cup rice
1⅓ cups water
Pinch or two of salt

Rinse off the rice once or more until the rinse water is clear. Put the water in a pot with the white rice. Add the pinch of salt, cover, and heat to boiling. Then reduce heat to simmer and cook until tender, about 15 minutes. It is best not to remove the lid from the pot while the rice is cooking for two reasons: Moisture is lost in the escaping steam and heat and moderate pressure will be dissipated. You can tell when the rice is done by the pleasing aroma and by the lack of the bubbling sound of the water – it's evaporated.

Tofu with Mushrooms, Carrot, and Spinach

To make this dish you do not really need all the ingredients, but the colors are more decorative if you use them all. Tofu usually comes in two varieties: firm and soft. For this recipe the firm tofu is best, as the soft will crumble and fall apart when stir-fried.

Serves 4 to 6 people

1 tablespoon olive oil

1 small red onion, sliced

2 cloves garlic, minced

1 tablespoon fresh ginger, grated

1 large carrot, cut into ovals or ovals cut into strips

Salt

6 to 8 mushrooms, sliced

8 ounces tofu, cut into cubes

6 to 8 ounces spinach or chard, cut into 1-inch pieces

Soy sauce

Heat the olive oil in a skillet and then sauté the onion for a couple minutes. Add the garlic, ginger, carrot, and a sprinkling of salt. Reduce heat and continue cooking for another 2 to 3 minutes. Add the mushrooms and stir to combine, then add a couple spoonfuls of water, cover, and cook over low heat until the carrots are soft. Add the tofu and spinach, cover, and cook to heat and wilt the spinach. Season lightly with soy sauce.

Struggling to Get It Right and Eating the Results

Potato Fiascoes

When I was the head cook at the Tassajara Zen community in the late sixties we ate a lot of brown rice. About a third of the residents were followers of "Zen" macrobiotics who believed that short-grained brown rice was the perfect food. Chewing each mouthful of rice fifty to one hundred times was considered a meditation in and of itself.

While macrobiotics actually has nothing to do with Zen, and while I was not an adherent of this diet, I was obliged to cater to those who were, as they were loud and impassioned about eating the correct foods. They used to say that if one followed the proper diet, then one would feel peaceful and happy. Apparently this was true, because when they didn't have their proper food, they were outraged. Eventually our eating habits broadened, but meanwhile we lived under the tyranny of this diet plan. Twice during the brown rice era I tried cooking potatoes, and both times were failures. What did I know about cooking potatoes?

Once for the last dinner of a sesshin, our week of intensive meditation, I thought I would cook something "special," namely potatoes. After three months of brown rice, the thought of potatoes was pretty exciting. I could imagine just how delicious those baked potatoes would be, especially with some butter and sour cream. Nothing to it, right? You just put them into the oven and bake. Simple.

I wanted very much to delight everyone who had been diligently practicing meditation, to treat everyone to what in the eyes of our macrobiotic contingent was a forbidden fruit. "Deadly nightshade family," they would say, as if that explained why potatoes were unacceptable. Also I had heard that our teacher Suzuki Roshi loved potatoes. Unfortunately, the great feast turned into the great fiasco.

We filled the ovens with potatoes. I think we'd washed them and rubbed butter on them. I was thrilled. The potatoes were baking, and I thought that one and a half or two hours was plenty of time, but I found out it wasn't. Twenty minutes or so before mealtime, we opened the ovens to check and realized the potatoes were not getting baked. My spirits dropped. We turned the ovens up full blast, but it was too little too late.

Under everyday circumstances one simply apologizes and delays the meal. In the restaurant business one might offer complimentary wine, but in the Zen tradition, when the bell rings, the food is served. Period. Meals are never late. No excuses, no delays. Certainly for the meditators it is quite reassuring. When the time comes for dinner, dinner comes. The cooks, especially the head cook, take on all the pressure and anxiety of making sure the food is ready. You do your best under the circumstances.

I have since discovered that in certain other spiritual traditions, when the food is ready, the bell is rung. This way the anxiety is shared. "So the food's not ready. That's not a problem for you, is it? You'll hear the bell when it's ready." But this was Zen.

The potatoes never got baked. I had neglected to take into account how much an oven cools off when a large quantity of food is added. It can take an hour just to get the food up to oven temperature, let alone cook it. Also I have since learned that ovens heated with propane don't get as hot as those with natural gas or electricity. I kept thinking that maybe the potatoes wouldn't be that bad, but they were.

When we had finished cooking, we would put on clean aprons at the last minute and go to the meditation hall to serve the meal. Sweaty, frazzled, and fried from cooking, we would try to calm down, to settle and collect ourselves in the few moments before we would be entering the meditation space.

During serve-up the anticipation built up. Glorious baked potatoes. Eager eyes and noses prompted the thrusting of eager bowls to receive generous helpings of potatoes. With serve-up complete, dining commenced. For our meals in the meditation hall we use a set of eating bowls that comes with a spoon and chopsticks, and from the back of the meditation hall I watched with fascination and dread as Suzuki Roshi's spoon bounced off his potato.

The man was intrepid though, and he had done plenty of stonework in his time, so he proceeded to make a row of holes with his chopstick, and then chisel off a piece of potato using the end of his spoon. He did what he could to eat his potato.

As this comic scene unfolded, I was devastated, yet I had to carry on. I don't recall what happened to all those hunks of potato, if they were eaten, collected, or disposed of with the wash water. We had pots to wash and the kitchen to clean before I could fall apart. But I surprised myself by not falling apart all that much, because I knew that I had made a good effort.

Sometime later I apologized to my teacher Suzuki Roshi, who shrugged. He always made me feel good about myself. Potatoes might mock my effort, but he knew my heart.

The next sesshin I decided to try again. This time I planned to serve mashed potatoes, and get them cooked well ahead of time. Only when we were putting the potatoes into the wooden rice buckets for serve-up did I begin to sense that a different kind of disaster was in store. The buckets just weren't very full. Once again I had made a miscalculation. We didn't have nearly enough.

Potatoes, it turns out, are like eggs. When they are mashed they don't look like much. People who will eat only one or two eggs fried, poached, or hard-boiled, will happily indulge in a three-egg omelet, and if the eggs are scrambled, some people will eat four or five. A big potato doesn't look like much when it is mashed.

The meditation hall was buzzing. Those being served first insisted on having their bowls heaping full of potatoes. The servers were trying to conserve their supply of potato, and in silence would anxiously glance down the row at all the people who remained to be served, attempting to indicate their dilemma to the early recipients, but it didn't work. Those at the ends of the rows got little spoonfuls.

The hall swarmed with emotion: bitterness, greed, joy, passion, resentment.

The potatoes wreaked havoc. I think we heated up some leftover brown rice for seconds, but the damage had been done. When it comes to food, people want a full bowl, by golly, regardless of how little that leaves for others a few steps away. Another low point to add to the highlights of my cooking career.

Recipes

In the recipes that follow I am not going to tell you how to make baked or mashed potatoes, but instead I would like to offer you some of my other favorite ways to prepare potatoes. Though I have certainly tried any number of different potatoes – Russet, rose-fir, yellow Finn, Yukon gold, and the purple Andean potatoes – I keep coming back to red potatoes. I like them, their redness, their versatility, and their smooth, creamy texture.

Potatoes Roasted with Garlic and Rosemary
Oven-Fried Potatoes
Potato Salad with Arugula and Garlic-Mustard Vinaigrette

Potatoes Roasted with Garlic and Rosemary

This is my version of what has become a nouveau "classic" potato dish.
 Serves 4 to 6 people

2 pounds red potatoes

1 head of garlic

4 to 6 4-inch sprigs of fresh rosemary *or* summer savory

Olive oil

Salt

Black pepper

Wash the potatoes and cut them into large chunks. The smallest ones (golf-ball size) may be left whole. Otherwise cut them into halves or quarters. Peel the garlic and cut the largest cloves in halves or thirds. Cut the rosemary into 2-inch pieces. Combine the potatoes, garlic, and rosemary in a bowl. Toss with 1 to 2 tablespoons of olive oil, then place in an oiled baking dish. Sprinkle with salt and pepper, cover with a lid or aluminum foil, and bake 45 minutes to an hour in a 400-degree oven.

Oven-Fried Potatoes

This is as close as I come to making French fries, because I almost never fill a pan with fat and fry in it. This method is much more relaxing and takes less attention than deep-frying. Compared to the roasted potatoes above, these will come out dry and crisp rather than moist and creamy.

Serves 4 people

2 pounds red potatoes
2 tablespoons olive oil
Salt
A few drops of cider vinegar

Slice the red potatoes into rounds about a ¼ inch thick, then into strips – like French fries! Place in a bowl and toss with the olive oil to coat. Spread out on one or more baking sheets, sprinkle with salt and bake in a fairly hot (425 to 450-degree) oven. Bake for 45 minutes to 1 hour, turning every 15 to 20 minutes, so that they become browned on various sides. Longer baking will tend to brown the outside of the potatoes, but too long and they will become dried out. Remove from oven and serve, sprinkled with more salt and perhaps a few drops of cider vinegar.

Potato Salad with Arugula and Garlic-Mustard Vinaigrette

I make this salad over and over for lunches, dinners, and picnics. The arugula, with its mustardy quality, is an enjoyable green, leafy complement to the potato, but if arugula is not available, spinach is a good alternative.

Serves 4 to 6 people

2 pounds red potatoes
4 large shallots, thinly sliced
4 cloves garlic, minced
2 tablespoons Dijon mustard
½ teaspoon salt
Freshly ground pepper
3 tablespoons sherry wine (or balsamic) vinegar
½ cup olive oil
½ bunch arugula or small spinach

Wash the potatoes and cut into bite-size chunks: halves, quarters, eighths, or more, depending on the size of the potatoes. Cook potatoes in boiling, salted water 6 to 8 minutes until tender. (Personally I am not a fan of those *al dente* potato salads.)

Whisk shallots, garlic, mustard, salt and several grinds of pepper together with the vinegar in large bowl. Whisk in the olive oil.

Drain the potatoes. Toss with the vinaigrette. Cool for 20 to 30 minutes.

Cut the stems off the arugula and if the leaves are large, cut them in half. (You can cut the whole bunch while it is nicely bound together, then wash and spin dry what you will use.) Fold the arugula into the potatoes. Depending on how hot the potatoes still are, the arugula may cook slightly, soften, and sweeten. The dish can sit a while before serving, if desired. Adjust salt, pepper, and vinegar to taste before serving.

Rotten Pickles

Suzuki Roshi once told us a story from his childhood that left a particularly poignant taste in my mouth. Food is not just food. The entire universe comes along with it. Human nature makes its appearance bite after bite.

As a boy of perhaps ten or eleven, Suzuki Roshi had been sent by his father to study with another Zen teacher who was his father's disciple. There were apparently four or five boys altogether. In the spring they would help their teacher make daikon pickles. The long white radishes would be layered in barrels with salt and nuka (rice bran).

We used to make these pickles at Tassajara. The mixture is dry at the outset, but as the barrel sits, water comes out of the radishes, moistening the nuka and thereby salting the radishes. At least that's how it's meant to work. The salt acts as a preservative. The rice bran provides flavor and perhaps nutrition.

One year at the temple in Japan a batch of pickles the boys and their teacher made didn't quite make it; a number of the radishes developed noticeably "off" flavors, which happens when there is not enough salt. What to do when something doesn't turn out the way it should, the way you wanted, the way you planned? The teacher served them anyway! All well and good for him, but boys will be boys, and the young Suzuki Roshi and his companions refused to eat them. Each day the pickles would be served, and each day studiously avoided.

At last Suzuki Roshi decided to take matters into his own hands. One night he got the pickles, took them out to the far end of the garden, dug a hole, and buried them. Isn't that what you do with something distasteful? Dig a hole, put the rotten stuff in, and cover it with dirt – a straightforward, elegant solution, returning earth to earth. Let it compost. Keep it covered.

Yet life is not always that simple. The next day the pickles were back on the table! Things you bury don't always stay put. What an unpleasant surprise, and what a sinking feeling to have what you were trying to hide come out into the open. The teacher, however, did not say anything about the pickles having been buried, or whether or not he knew who buried them. He merely stated that those pickles would have to be eaten before they got anything else to eat.

Sometimes we have no choice; we have to taste and digest what we find distasteful. Suzuki Roshi said that it was his first experience of "no-thought," when the conceptualizing mind stops and one experiences something nonreactively with no added comments. Chew and swallow. Chew and swallow. He could only eat the pickles if his mind did not produce a single thought.

The world itself is swallowed up. For a time the storyline disappears. No more "This is awful," "How distasteful," "How unfair," "What did I do to deserve this," or even "Yuck," because then you would have to spit the pickle out, or choke it up. Just chew and swallow.

We need to be able to conceptualize, to decide what is good to eat and what is not, yet we can suffer a lot by trying to have nothing but delicious experiences. Inevitably we will have to chew on and digest some difficult, painful moments.

We would like to say, "Skip the pickles," but this is the great dilemma that life serves up: Not everything is tasty and cooked to perfection and there is no way to avoid all that is unpleasant. If we become too finicky, we just don't eat.

The dirt of our life contains both good and bad, sweet and pungent. The cook unearths what is there, and labors to make it nourishing.

Recipes

Leftovers are something that can turn nasty and distasteful if left too long. Over the years I have become more careful about distinguishing between what is still edible and what is "gone." Here are some ways to work with the remains that are still worth eating.

You can study for yourself what you like and what you don't, what works for you and what doesn't. The various ingredients need to be kept in proportion.

Leftover cereal, for instance, can be incorporated into minestrone or one of the leftover grain dishes, as long as it is not the primary ingredient. Most leftovers are like this.

Working with Leftovers

Making Minestrone
Fried Rice
Burrito with Leftovers
Omelet with Leftovers

Making Minestrone

I make my minestrone without beef or veal or chicken stock. I think of it as a vegetable soup distinguished by the presence of sautéed onions and garlic, with seasoning by oregano and garnish by Parmesan, and, for goodness sakes, buy yourself a chunk of cheese and grate it at home.

Few foods are more distasteful to me than the ersatz cheese powder that comes in a cardboard container. What will people pay not to grate cheese? Three to five times the price, at least, because "grating cheese is so hard, and I've grown accustomed to that stale flavor with hints of refrigerator odor."

Serves 4 to 6 people

1 cup cooked beans or bean soup (kidney, white, or lentil or perhaps black, pinto, or garbanzo)
2 to 4 cups mixed leftovers (Don't try to think these up, open your refrigerator: pasta, potatoes, cereal, vegetables, green salads, bread, old dry crusts of cheese.)

1 yellow onion, diced

1 to 2 tablespoons olive oil

2 to 4 cloves garlic, minced

½ teaspoon (or more) dried thyme

1 teaspoon (or more) dried oregano

Tomato paste (optional)

Soy sauce (optional)

Salt

Pepper, black or red

Flat-leaf parsley, minced (optional)

Freshly grated Parmesan *or* Asiago cheese

Start by getting all those little containers out of the refrigerator. See what's in them. Discriminate. Those with visible growth get tossed. Apply the nose test as well to make sure the contents have not soured and are not emitting telltale airplane glue aromas. Why don't we do this more often?

Keep in mind that the quantities listed are approximations, but the beans will help to "beef up" the soup. Leftover cereal, including wheat, corn, or oatmeal, will give the soup more body. Prep the other leftovers by cutting the pasta, potatoes, vegetables, or bread into spoon-size pieces or, if you prefer, by blending them, which is especially good for any green salad. Perhaps some blended and some whole. Place leftovers in a pot with water to cover, and begin heating.

Sauté the onion in a hot skillet with the olive oil. After a minute or so, add the garlic and dried herbs. Cook another couple of minutes, then add to the soup. Rinse out the skillet with a small amount of water and add it to the soup.

Check for color and seasoning. If pale-colored or watery, add some tomato paste, soy sauce, or both. Season with salt and red or black pepper, or more herbs. The minced parsley added at the end will add color and bring up the flavors. Garnish with the grated cheese.

Fried Rice

Sautéed onion and garlic will help most any leftovers. The old saw used to be that if the housewife had not done any dinner prep all day, she was advised to have onions sautéing when her husband got home. Here fresh ginger is used as well.

In this recipe the vegetables are fried, and not the rice, which is steamed. Still it goes by the common name of "fried rice."

Serves 4 to 6 people

1 yellow onion, diced *or* 1 leek (white and pale green parts)

1 to 2 tablespoons sesame or olive oil

2 stalks celery, diced

1 carrot, diced *or* some leftover vegetables, diced

2 to 4 cloves garlic, minced

1 to 2 tablespoons fresh ginger, grated

About 2 cups cooked rice, brown or white (or other grain: groats, millet, cracked wheat)

Soy sauce

Salt

Pepper, black or red

1 or 2 eggs (optional)

3 to 5 green onions, thinly sliced, for garnish

Slice the onion or leek and then sauté it in hot oil for a minute or two. Add the celery and carrot and continue to cook, stirring. When the celery and carrot begin to soften add the garlic and ginger, and cook for another minute. If using leftover vegetables, cook the garlic and ginger with the onions, then add the vegetables. If the vegetables begin to stick to the pot add a bit of water to loosen.

Add the rice, and stir to combine. Add 2 to 4 tablespoons of water, cover, and heat over low flame heat – a few minutes will do. Season with soy sauce, salt, and pepper – black or cayenne, or Tabasco or chili. If adding egg, beat first, then stir in and cook briefly before serving. It will pretty much disappear but gives the dish a heartier feeling. Garnish with the green onions.

Burrito with Leftovers

This is such an easy, delicious, and quick way to make a meal. Take some leftover beans, grains, and/or vegetables and add some green chilies or salsa, some avocado or cheese.

Makes 1 burrito

1 cup leftovers (grains, beans, cooked vegetables or vegetable salads, pastas? potatoes?)
1 flour tortilla

Optional:
Choice of chili, such as fresh green chili, minced; dried red chili powder
    (see page 19), *or* Ortega canned green chilies, diced
Salsa
Avocado (optional)
¼ cup grated cheese, (any kind, whatever's handy)

If there are large pieces of vegetable among the leftovers, you can always chop them up somewhat. Smaller pieces will mix with the other ingredients more easily.

Heat up the leftovers you'll be using in a small (preferably nonstick) skillet, covered with a lid, over moderate heat. If the leftovers are dry, add a spoonful or two of water, so that the ingredients don't stick to the bottom of the pan while they are steam-heating.

Depending on what you have around or might like, spice up the leftovers with your choice of chili, salsa, or both.

In a separate skillet you can be heating the tortilla. Sometimes I sprinkle the cheese directly onto the tortilla so that it melts while the tortilla is heating. Then put the tortilla on the counter and the filling across the middle, leaving a bit of space at both ends. Add the avocado here if you are using it. Pick up the near and far edge of the tortilla, and rock the ingredients into a log shape across the middle. Then release those edges, and with the tortilla out flat, fold the right and left ends of the tortilla up and slightly over the filling. Then roll up the burrito starting at the near edge.

If you are making several, set them in a baking dish or skillet to reheat, if necessary. Otherwise, arrange on a platter or individual plates for serve-up.

Omelet with Leftovers

Perhaps you have your favorite omelet, or perhaps you are no longer eating eggs, taking the word of modern science, which says no, no eggs. Butter for some time was a no, and now may or may not be any worse than margarine with its machine-made fats. Oh, how will we ever keep track? And hey, did you know? Now we can eat walnuts again – they reduce cholesterol.

Yet omelets with leftovers are often quite good and, shifting paradigms here, in Chinese medicine, eggs provide Essence, which, according to that terminology, is the fundamental basis of nutrition. I've made omelets with fillings from pasta dishes, spaghetti, or vegetables. This recipe fills you in on the details and gives you some seasoning options.

For 4 people

1 to 2 cups leftovers (pastas, potatoes, or vegetables)
6 green onions, thinly sliced
2 tablespoons butter
2 teaspoons fresh thyme *or* ¼ cup chopped fresh basil, *or* ½ teaspoon of either
     herb, dried
6 eggs, beaten
Salt and pepper
Grated cheese (what kind do you have?)

Check out your leftovers. For this dish I would avoid using leftover cereals (which I have found suitable for some of the other leftover dishes) or dishes with fruit. As in the previous recipes, cut up the leftovers so everything will blend together. Do you have anything with mushrooms or tomatoes? They go especially well with eggs.

Begin by heating the green onion with half the butter. Once it softens, add the leftovers, cover, and heat. Add a spoonful or two of water if necessary to keep them from sticking. After heating, drain off any excess liquid and combine the liquid with the beaten eggs. If using dried herbs add them to the eggs, and season with salt and pepper as well.

Remove the leftovers from the pan and clean it out. Melt the remaining tablespoon of butter and have the pan moderately hot before adding the eggs. Then reduce the heat and cook over low heat until the eggs thicken. If necessary lift up some of the cooked egg, and tilt the pan so that more of the liquid egg runs underneath.

When the eggs are nearly cooked, put the heated vegetables on them. Fold the egg over the vegetables or leave it open-faced, and top with the cheese. Garnish with the fresh herbs, if you are using them.

Unearthing Greed

When I started zen practice I maintained an inflated level of self-esteem by projecting all my pain and nastiness onto other people. *They* were greedy, *they* were mean, *they* were angry. I thought I didn't have problems like that: I was a zen student; I was spiritual. I was getting enlightened. It turned out that what I was getting enlightened about was my own delusion.

One of the first things I discovered was greed. I would watch other people at teatime strategically position themselves to move in for the biggest pieces once the chanting was concluded.

"Me? I'm not greedy like that," I would observe from a distance, "I can wait. Look at how greedy *they* are. Even if I was that greedy, I wouldn't be so obvious about it." No, certainly not, I had it pretty well hidden, even from myself. Greedy? No, I was going to be an outstanding Zen student, better than the others.

Suzuki Roshi's words wouldn't catch up with me until later: "To be better or worse than other students is not the point. As long as you are comparing yourself to others, you are not practicing our way. Your practice may be quite good, but when you stop to compare, you lose the way."

Yet try as I might I could not keep all my faults hidden. Once Mrs. Suzuki, the Roshi's wife, gave me a big box of mixed salted nuts. I was elated. What a sign of my good practice! I kept them sitting around my room for some time, their presence reassuring me that I was chosen. I assumed that I would find some occasion to share them with others, but then the chink in my armor of denial appeared, not obviously, but slowly and insidiously.

In the silence and aloneness of my room I opened the box of nuts and ate a few. I savored them, and found them exquisitely delicious. Still, I thought, I would share them – tomorrow. Yet day after day went by, and each afternoon I would sneak back to my room and eat a few nuts, while day after day it became more and more apparent that I had no intention of sharing them. I felt ensnared, helpless. "No," I realized, "I am not going to share them."

I felt petty, yet the nuts were excruciatingly tasty. I wasn't going to give up even a single second of that delight. I ate them secretly and intently, savoring each one. I'm not sure the experience could be called pleasurable, but it was certainly riveting. After more than two weeks, the box of nuts was gone, and I had to admit that was greed, I am as greedy as the next person.

Although I hadn't wanted to see how greedy I was, the evidence was too plain. I couldn't really say that those nuts were mine, that they belonged to me, and that I could do with them what I wanted. They had been a gift, and I had appropriated the gift all for myself, instead of sharing it. To share the gift would have given others some of the same satisfaction and delight I had felt initially, before my stinginess turned it into something cold and banal.

Suzuki Roshi had said that the point of Zen is to own your own body and mind. I was only beginning to realize that meant owning up to my shortcomings.

Recipes

I find that offering others my good wishes helps to loosen greed's grip on me. May you be happy, healthy, and free from clinging, hate, and delusion. May you grow in wisdom and compassion, and enjoy ease of well-being. May this food nourish you – in body, mind, and spirit.

Potato Leek Soup
Green Bean and Tomato Salad with Feta Cheese

Potato Leek Soup

This soup uses sautéed bell peppers to enhance its basic flavors.
Serves 4 to 6 people

1 pound potatoes, cubed

2 cups water

2 leeks (about 12 ounces), white and pale green parts only

1 teaspoon dried thyme

¼ teaspoon salt

A few twists of freshly ground black pepper

¼ cup flat-leaf parsley, minced

½ green bell pepper (about 3 ounces)

½ red bell pepper (about 3 ounces)

1 tablespoon olive oil

4 cloves garlic, minced

1 tablespoon white wine vinegar

¼ cup fresh basil or tarragon, finely sliced for garnish, if available

Start the potato cubes heating in water. Slice leeks, using white and pale green portions, and wash to remove dirt beneath the surface. Add to the potatoes along with thyme, salt, pepper, and parsley. Cook 30 to 40 minutes until vegetables are soft.

Cut pepper halves lengthwise in thirds, then crosswise into thin strips. Sauté for 2 to 3 minutes in olive oil. Add the garlic, mix in, and continue cooking another minute or two. Add the vinegar. Cover, reduce heat, and cook until tender. Set aside.

Put at least some of the soup through a chinois or strainer, or pulverize with a masher. (Do not use a blender, as this will make the soup gummy.) Add water to desired thickness. Add the pepper mixture and adjust seasoning.

Garnish with fresh herbs, if available, or use a sprinkling of dried.

Green Bean and Tomato Salad with Feta Cheese

For many years big tough green beans called Kentucky Wonder beans were all that was available at supermarkets. Now more tender varieties such as Blue Lake are likely to be carried there.

The back end of vegetable peelers used to have a slot that you could pull green beans through to shred them lengthwise, one by one. That was fun – and made Kentucky Wonders edible. Now I still like to cut the green beans lengthwise, but I do it by hand with my Japanese vegetable knife.

Serves 4 people

½ pound green beans

1½ pounds flavorful ripe tomatoes

½ pound feta cheese

2 shallots, finely diced or minced

2 to 3 tablespoons olive oil

1 tablespoon red wine vinegar

3 to 4 pinches of salt

Black pepper

⅓ cup fresh basil, cut into thin strips *or* 2 to 3 teaspoons fresh thyme or
   marjoram, minced

After washing and trimming off the ends of the green beans – you need only to cut off the stem end – cut the beans in half lengthwise or in 3-inch-long diagonal strips. This is certainly not an exact procedure as the beans curve in various directions, but it will allow them to cook more quickly and become more tender.

Steam or blanch the beans until they are tender, about 4 to 5 minutes. Drain.

Cut the tomatoes in half lengthwise, cut out the core, and then cut into wedges. Cut the feta cheese into cubes or strips.

Toss the green beans, tomatoes, cheese, and shallots together with the olive oil, then with the vinegar. Before adding salt, check to see how salty the feta cheese is – I get a kind that is fairly mild – then season to taste with salt and pepper.

Garnish with the fresh herbs.

Coming to Your Senses

Back in the sixties at Tassajara our diet was fairly austere. Nowadays we have a "back door café," where we put out fruit, breads, jams, peanut butter, and other leftovers for snacking, but in those days

Struggling to Get It Right and Eating the Results

our hunger was focused on the three meals. So these few occasions to eat took on great significance. Some of us ate voluminously and ravenously. Especially at lunch a feeding frenzy would often unfold.

A group of us would eat sixteen, eighteen, twenty half slices of bread, the equivalent of eight to ten full slices, and this was not light and airy but home-made, chewy, dense bread, plus gobs of spread. And very few people gained weight. Perhaps all those calories got burned up in the frenzy to eat more, though I was working pretty long hours, too.

All this eating was done in just a few scant minutes. In Zen practice apparently it was not appropriate to savor food or linger over it. Within five or six minutes after we had finished our pre-meal chanting and begun eating, seconds would be served, so after an initial taste of each bowl, a quick decision was needed. "Which bowl do I want more of the most? Oh, oh, here come the servers. Stuff it in." It was painful to be so driven.

I noticed several things even in those times of seemingly insatiable appetite. Initially during a meal I would be aware of the flavor and texture of foods – the creamy nuttiness of oatmeal, the crunch and earthiness of carrot – and with this experiencing of the food a wonderfully sweet pleasure arose. Yet as soon as I decided, "I want more of that!" the pleasure ceased; the flavor and texture disappeared. All that remained was craving, a focus on "getting more" (receiving seconds), even though I already had "more" right there in my bowl (which I had to get rid of in order to get more).

Since we were sitting cross-legged for our meals, my sore legs were also a pivotal factor. Being absorbed by eating meant that I would be less preoccupied with my aching knees and painful legs, so if anything, I wished that the meal would go faster. I wished that the pain would go away, that this would all be over with. Having food in my bowls and in my mouth to occupy my awareness seemed like a useful way to take my mind off the pain. Isn't that the reason to overeat? To make the pain go away?

Looking back at this I am reminded of James Baraz, a vipassana teacher, describing his infant son eating strawberries: If he couldn't have both hands full of strawberries while eating strawberries he would start screaming in frustration, even though you could see his mouth was full of strawberry.

A little awareness is such a difficult thing. You see what a fool you are being and continue helplessly in the grip of the same foolishness, but the awareness does not go away. What an embarrassment.

At some point I took a simple, yet momentous step against the current: I would just eat. I would just taste and experience each mouthful, setting aside all considerations of the future and whether or not it would bring more of the same or not. When the meal was over, I would have *eaten* instead of having chased after imagined delights overlooking what was already in my mouth.

Overnight I started eating half as much and feeling more satisfaction than ever. I learned to ignore all of my scattered-brained objections: "But this is so dumb and boring," "How will I get more if I don't rush through what I have?" "Where's the fun and excitement of chasing after things?" Still I knew that I didn't want to end up being at the mercy of my desire, missing out on the pleasures of root, shoot, leaf, and fruit.

Come to your senses. It is not the things of this world, be they chocolate or brown rice, that lead you astray. Losing your way comes from giving no mind to what is present while chasing after imaginary pleasures which are illusive and unobtainable. To wake up is to know what is already yours.

Recipes

Here are some dishes that bring me down to earth and help me come to my senses: wholesome food with a gentle and tasty touch of spiciness.

Spanish Rice
Refried Pinto Beans
Cilantro Relish

Spanish Rice

Serves 4 people modestly

1 cup long-grain white rice

½ medium red onion (about 3 ounces), diced

1 tablespoon olive oil

½ green bell pepper, diced

1 clove garlic, minced

¼ teaspoon salt

1-pound can whole tomato (plain), with its liquid

Roast the rice in a dry skillet over moderate heat, stirring as needed, until it appears toasted and is fragrant.

Sauté the onion in olive oil for a minute, then add the green pepper, garlic, and salt. Cook another minute or two.

Coarsely chop the canned tomato and, if necessary, add water to make 2 cups. Add it to the onions and peppers and stir to get the juices off the bottom of the pan. Combine this mixture with the rice, and cook in a covered pot about 15 minutes until tender. Open the pot and stir, then cover and let sit a few minutes before serving.

Refried Pinto Beans

Serves 4 people modestly

1 cup pinto beans, soaked overnight, if possible

4 cups water

1 tablespoon dried oregano

1 medium yellow onion (about 8 ounces), diced

1 tablespoon olive oil

4 cloves garlic, minced

1 to 2 tablespoons chili powder

2 teaspoons cumin seed, freshly ground

½ teaspoon salt

½ cup plain yogurt

4 ounces Jack cheese (or equivalent), grated

1 teaspoon red wine vinegar

Cook the pinto beans in water with the dried oregano. If you have soaked the beans they will cook in about half the time, perhaps 45 to 50 minutes. With a pressure cooker the beans will cook in 20 to 25 minutes. Without a pressure cooker and without soaking, they will take a good hour and a half. As the beans cook, make sure they remain covered with water.

Sauté the onion in the oil for a minute or two. Add the garlic, chili, cumin, and salt, and continue cooking another minute or so. Set aside.

When the bean are soft, drain them, and puree with a hand blender, adding back as much liquid as you need to give the beans a thick, creamy consistency. Add the onion and seasonings, and cook a few minutes over moderately low heat.

Stir in the yogurt and cheese. Add the vinegar and then check the seasoning, especially for salt, vinegar, oregano, and degree of pungency. You can always add more chili powder or a pinch or two of cayenne pepper to make it hotter.

Cilantro Relish

This is intended as an accompaniment for the Spanish Rice and the Refried Pinto Beans. Serve on the side and let people use it to garnish either or both dishes.

Makes ¾ to 1 cup

1 bunch of cilantro

2 shallots, minced

1 clove garlic, minced

2 tablespoons lemon juice

Pinch of salt
Pinch of sugar

Rinse off the cilantro and spin it dry. I try to keep the bunch together, so that it can be easily minced. Roll the bunch into a log shape and cut off narrow pieces. Or cut the cilantro into 1-inch pieces and put in food processor.

Combine the cilantro with the other ingredients and adjust seasoning.

Anger Appears Unannounced

Before I became a cook or a zen student I rarely thought of myself as an uptight person. I was sincere, conscientious, and laid-back. More than others. When anger appeared unannounced, that wasn't me. That was something I couldn't help, and I wasn't going to have anything to do with it.

Everybody else knew how angry I could be. One of life's great ironies is that everybody else already knows the very thing you are trying to hide from them, and all you succeed in doing is hiding it from yourself.

Even though I wasn't going to have anything to do with anger, I began being angry more and more often. I was angry to have to wake up, angry to have to get up and wash up, angry to have to dress. Moment after moment was distasteful, disgusting, objectionable: fatigue, painful knees, tedious labor, people (too slow, too greedy, too talkative, too sullen), things (out of place, demanding attention).

In a zen parable this is the person who wakes up with a piece of shit on his nose. Throughout the day everyone he meets "stinks" and everything he does "stinks." Many years may pass before he realizes the shit is on his own nose, and he must "wash his face."

I hadn't yet noticed the shit on my nose, but people walked up to me as though they were looking around the corner of a building . . . just to check and see what mood I was in before they got close. I guess they could smell it.

I threw fits and sometimes I threw things: spoons, spatulas, other kitchen utensils. Things can easily be exaggerated, and there are stories that I threw knives, but, no, it was only handy lightweight objects, including my glasses. They would get so steamed up and greasy, and became impossible to see through. A few fits like that, and I stopped wearing glasses: Skip it, if they don't know how to stay clean. I just left them behind the stove where they had dropped after hitting the wall.

Not that my fits ever made a difference. Expressing anger is often like unleashing a ferocious, penned-up bull: a lot of damage and precious little communication. If I attacked the way people worked – "Can't you pay more attention to what you are doing?" – nobody seemed to notice *what* I said. All they noticed was the anger – "Gee, you're really mad, aren't you?" Anger was getting in the way. It wasn't helping.

At last I found myself wondering, "How am I ever going to find out what to do with anger, if I never have anything to do with it? If I want to find out what to do with anger, I better begin studying how it works." Once I started observing, I noticed pretty quickly that the object of my anger was not the cause of anger: The shit had been on my nose. As far as I could tell anger was simply a reaction to the fact that the world did not behave the way I wanted it to, the way I thought it should. Apparently the universe is not set up according to my agenda.

Experiencing anger is so painful that we would rather suppress it or direct it at someone else as a way of keeping our distance from it. Often we look for a quick fix: "How do I get rid of my anger without ever having to relate to it?" Yet deciding to get acquainted with anger is pivotal. That begins a lengthy process akin to becoming intimate with a person and finding out how to live with him or her.

Becoming companions with your anger is a way to transform it. When you can tolerate having your anger around, it is not so intolerable. Anger loses its grip on you.

Struggling to Get It Right and Eating the Results

Recipes

Getting to know your ingredients is just as important as becoming acquainted with your emotions. After a while you find out how to work with things. This menu follows a pattern I've become fond of having: olive oil bread, a soup, and a salad. Depending on the season different ingredients will be utilized. In this case the soup is probably more of a fall-winter one and the salad more of a summer one.

First I start the bread, then the soup, and finally the salad.

Focaccia: Olive Oil Bread with Fresh Rosemary
Cauliflower Tomato Soup with Herbes de Provence
Corn Salad with Zucchini and Roasted Red Pepper

Focaccia: Olive Oil Bread with Fresh Rosemary

I walk out the door of my cottage along the uneven red brick path to the front gate, where the rosemary bush grows. I prune it so that I can continue to open the gate and walk by. Then I come in, mince the fresh herb, and begin the bread.

Often I omit the second rising, so that I can have the bread ready sooner. This dough can also be used for making pizza.

Makes 2 modest loaves, serves 4 to 6 people

2 cups warm water, under 125 degrees

2 tablespoons active dry yeast or 3 packets

3 tablespoons fresh rosemary, minced

4 to 6 tablespoons olive oil

2 teaspoons salt

1 cup unbleached white flour

1½ cups whole-wheat flour

3 cups unbleached white flour

Olive oil, for glazing

Coarse sea salt

Start with the water, making sure it is not too hot – it will feel just slightly warm on your hand. Stir in the yeast, then the rosemary, olive oil, and salt. Stir in the 1 cup of white flour and the whole-wheat flour. "Beat" about 100 strokes, which means stirring at the surface in short, quick, up-and-down strokes.

Fold in 2 cups of white flour, a half cup at a time. Turn the dough out on a floured board and knead for several minutes using up to another cup of flour to keep the dough from sticking. Knead until the dough is smooth and elastic.

Let the dough rise for about 1 hour until it doubles in size. Punch down and let rise another 40 minutes.

To shape into loaves, first divide the dough in half. Shape each half into a ball. (They can be baked in this shape.) I like to flatten out the ball into a rectangle, then roll up into a log shape. Then I flatten out the log and make parallel diagonal cuts crosswise, about an inch apart, leaving the sides attached, but cutting all the way through in the middle. I pull the ends lengthwise, so that the cuts are stretched into openings, forming a "ladder" shape.

Place on an oiled sheet pan, brush the top with olive oil, and sprinkle with coarse sea salt. (This makes it somewhat reminiscent of a soft pretzel.) Preheat oven to 375 degrees.

Let rise about 20 minutes, and then bake about 25 to 30 minutes until browned top and bottom.

Cauliflower Tomato Soup with Herbes de Provence

My love for vegetarian soups deepened when I was working at Greens Restaurant, where over the years we made so many excellent soups. Mostly I prefer my veg-

etable soups without any milk or cream, as then their flavor more clearly expresses the ingredients used.

Herbes de Provence *includes thyme, fennel, rosemary, lavender, and ?????*
Serves 4 to 6 people

2 cans (1 pound, 12 ounce each) whole peeled tomatoes or tomato pieces
2 tablespoon olive oil
1 yellow onion, diced
2 stalks celery, diced
2 cloves garlic, minced
2 to 3 teaspoons *herbes de Provence*
Salt
Black pepper
1 small to medium head of cauliflower
2 teaspoons dried oregano leaves
¼ cup parsley, minced *or* 2 green onions, thinly sliced, for garnish

Use plain canned tomatoes rather than one of the preseasoned varieties and blend them.

Heat a skillet, add a tablespoon of the olive oil and sauté the onion for a couple of minutes. Add the celery and cook another minute or two before adding the garlic, 2 teaspoons *herbes de Provence*, and a sprinkling of salt and pepper. Cook another minute, and then transfer to a soup pot with the tomatoes. Rinse out the skillet with a small amount of water and add it to the soup to retain the flavors from the bottom of the pan. Clean the skillet so that you can use it to cook the cauliflower.

Cut the cauliflower into spoon-size flowerettes. You can also use the stalk by cutting it in half vertically and then into narrow slices. Reheat the skillet, add the remaining tablespoon of olive oil, and start the cauliflower sautéing. After it has cooked for 3 to 4 minutes, add the oregano and a sprinkling of salt and pepper. Continue sautéing for another minute or two. Then add a couple of spoonfuls of water, cover, reduce the heat, and cook a few minutes until the cauliflower is just tender – I prefer it a bit "nutty."

Combine cauliflower with the tomato, check the seasoning, garnish, and serve.

Alternatively, you could sauté the cauliflower with the oregano, and then add it to the tomato. The cauliflower will take several minutes longer to become tender in that case.

Corn Salad
with Zucchini and Roasted Red Pepper

Serves 4 people

1 Roasted Red Pepper (see page 81)
2 zucchini
1 ear of fresh corn
Juice of 1 lemon
2 tablespoons honey
2 tablespoons fresh cilantro, minced
Fresh cilantro leaves, whole, for garnish

Cut the roasted peeled pepper into strips. Cut the zucchini lengthwise into quarters, then crosswise into 1-inch pieces. Cut the corn off the cob. Blanch the zucchini and corn briefly in boiling, lightly salted water. Remove after a minute or two and drain. Combine in bowl with the peppers. Dress with the lemon and honey mixed together and the minced cilantro. Garnish with some whole fresh cilantro leaves.

The Unwanted Guest Returns

Once I began to study anger I noticed that sometimes I could invite it to join in what I was doing – wash the dishes, scrub the pots, cut

Struggling to Get It Right and Eating the Results

the vegetables – and it brought a lot more spark than I was used to doing these things.

This didn't necessarily work though, because I often thought my anger was justifiable and appropriate. Where I had been saying that "I" was overcome with anger, that "I" couldn't help myself, I started owning that "I" am angry, and "I" have a right to be.

Doing this in a Zen monastery doesn't go over very well. Repressing anger may not be healthy, but expressing it all over others is not met with approval either. One resident said to me, "You can invite the visitor in, but you don't have to serve him tea." I couldn't understand the distinction. This visitor barged in unannounced, and I was trying to focus his energy a bit.

Suzuki Roshi spoke with me. When I confessed to not knowing what to do with anger, he responded rather slyly, "You can get angry if you want . . . but *don't*." I was startled. His timing had been impeccable. He had been so gentle, so understanding – "You can get angry if you want . . ." – then a pause, while I relaxed. "How nice," I thought, "he's sympathetic. What a relief, he's not telling me I can't get angry." Then that startling "don't," a "don't" that had such an unusual quality: not angry, not threatening, soft yet completely firm, rock solid, straight to the bottom. It stopped me cold, yet I wasn't angry or resentful.

Then after another pause, he was warm and friendly. "OK?" he beamed, as though it were no big deal. None of that "Is this clear, young man?" stuff. Just clearly, "I'm with you," – spiritual friend as much as a spiritual master. "OK," I agreed. If Suzuki Roshi said it, it must be possible.

But it wasn't easy. I kept getting angry, yet I also kept coming back to "You can get angry if you want, but don't." Maybe I was following the first part of the statement more than the second, because one day someone told me Katagiri Roshi wanted to see me. Katagiri was Suzuki Roshi's assistant for several years, and I had known him since he gave me meditation instruction the first day I arrived at Zen Center in San Francisco. Customarily the student requests the interview rather than the teacher. "Oh, oh," I thought with foreboding, what now? Although respectful, I think I also must have been a bit surly. "You asked to see me?"

Katagiri Roshi said yes, that my anger was a concern to the community, that I needed to do something about it so that I could live in peace and harmony with everyone. Everyone was quite concerned. I would have to do something.

"But," I protested, "if there is a problem with my anger, then other people will have to learn to live with it, won't they?"

"You must learn to control it," the Roshi replied.

"But isn't this Zen?" I demanded. "I'm just being sincere and honest about my feelings."

"Zen is to live in peace and harmony," was the Roshi's even response.

I continued to protest until finally the Roshi cut me off. "I'm giving you," he said with barely a trace of change in his demeanor or intonation, "a piece of advice." The room got very silent. His expression hung there. Take it or leave it? I felt no choice.

"OK, I'll work on it," I replied, "but how?"

Katagiri Roshi suggested that when I became angry while doing zazen, I could practice chanting to myself. My immediate response was to tell him that was not the correct way to practice zazen. Again he was quite patient with me. "Try chanting," he said.

The amazing thing was that it worked. Within two or three days I experienced a release from the pervasive feeling of tightness and rage. The practice of chanting or bowing, I have found, often works this way – to occupy the awareness so that it can let go of obsessive thoughts or emotions. Then when anger returns, it isn't as painfully afflictive.

Learning how to live with anger, how to use it, how to let go of it, is a more effective strategy than either exploding all over the place or always trying not to be angry. Shall we slice the cucumbers together?

Recipes

One guest I invite to return is wholesome food with bright colors and flavors. Shifting cultures here, I welcome you to try this North African-influenced menu. Though freshly grated Parmesan or Asiago cheese is not part of that culture, I still like to serve it on the side.

Couscous
Garbanzo Bean Stew with Spinach and Saffron
Cucumber Salad
Cucumber and Yogurt Salad

Couscous

Couscous is a grain product made from semolina wheat – sweet with a nutty flavor and easily chewable. Suzy Benghiat, in her book Middle Eastern Cooking, *says that "Some people – though certainly not North Africans – cook couscous like the rice for pilaf, but it does not give the lightness that real connoisseurs demand." It's a fine book, and I recommend it if you are in the demanding school (and even if you are not).*

Neither a North African nor a real connoisseur, I offer you the following recipe, which is the epitome of simple and quick. Couscous is available in packages at the supermarket or in bulk at natural food stores.

Most often the couscous is served with a stew of some sort, here one made with garbanzo beans.

1 cup couscous yields 3 average portions

1¼ cups water
Pinch of salt
1 cup couscous

Heat the water to boiling in a saucepot for which you have a tight-fitting lid. Add the salt, stir in the couscous, cover, and turn off the heat. Let sit 5 minutes, then stir or "fluff" with a fork, cover and let sit until serve-up – up to another 10 minutes or so.

Garbanzo Bean Stew with Spinach and Saffron

This stew has few ingredients and a delicate flavor. If you would like a more complex dish, consider adding any or all of the following: tomatoes, cumin seed, ground chili, oregano. See what you think.

Serves 4 to 6 people

1 cup garbanzo beans, soaked overnight or during the day, *or* a good deal longer cooking time

6 cups water

1 yellow onion, sliced

1 tablespoon olive oil

3 cloves garlic, minced

⅛ teaspoon saffron threads

Salt and pepper

1 bunch spinach (about 1 pound)

Rinse off the dried beans, and then soak them in 6 cups of water overnight or during the day. Cook in the same water until the beans are tender, about 1 hour or so. Bring the water to a boil, and then reduce the heat so that the water bubbles slightly. I leave a lid on the pot, somewhat ajar. Alternatively, the beans can be pressure-cooked for 30 minutes.

If you are like me, and have neglected to soak the beans, allow up to 2 hours for the beans to become tender. Check the beans periodically to make sure there is still enough water to keep the beans covered. Garbanzo beans seem to take longer to become tender than any other bean – I don't know why.

Sauté the onion in the olive oil in a large skillet until they are soft and trans-lucent. Add the garlic and cook another minute or two. Then put in the cooked beans along with the bean broth (unless it seems excessively soupy). Add the saffron threads and a mild amount of salt and black pepper. Let it stew.

With spinach I take a little extra time and cut off all the large stalks at the base of the large leaves, then I cut the off the root at the bottom of the stalks. Next I sort out any usable small leaves from among the stalks before discard-ing them.

Wash the spinach by immersing in plenty of water, and cut crosswise into 1-inch pieces. Add the spinach to the beans and onions, cover with a lid, and cook a minute or two until the spinach wilts. Check the seasoning and serve.

Cucumber Salad

I like cucumber simply sliced and then dressed with vinegar sweetened with sugar. To make it more colorful and appetizing, add a few orange slices, perhaps some chives or green onion. To give it something of a Middle Eastern flavor, a few drops of orange flower water or rose water could also be added.

Cucumber and Yogurt Salad

Cucumber combined with yogurt makes a cooling side-dish, and is used in several traditional cuisines. To give the salad a Middle Eastern flavor, season the sliced cucumber and yogurt with salt, pepper, and fresh chopped mint. For an Indian flavor mix the sliced cucumber and yogurt with salt and pepper, perhaps minced green chili, chopped cilantro, and/or ground cumin seeds.

An Egg Conspires

I'm amazed that little things can be so terribly upsetting sometimes. Those little things that only happen when I am incapable of dealing with them, when I am tired, grumpy, or preoccupied. This morning I am rushing to have breakfast and be on my way. Reaching for an egg, I find that it sticks tightly, impeccably, to its cubbyhole in the door of the refrigerator. Gentle repeated efforts fail to nudge it.

I caution myself not to break it while trying to free it, but with my first firm wiggle, the egg cracks. Slime, both clear and yellow, oozes out, heading like water for lower ground – the shelves below filled with their various jars of jam, mustard, soda, and things long forgotten. What a mess! (But it's so minor . . .) Shall I run out and get a sponge or paper towel, leaving the egg to drool? Or cup my hands to prevent immediate splattering?

A thought arises, "I can't stand it." Avoiding high cholesterol is not the only reason not to eat eggs.

Now, I've gotten accustomed to people being people, and to never knowing what stunt they'll pull next: disappear, flake out, flare up, cop out. But for *things* to pull such stunts, at times I find this seriously aggravating, especially when I could use some simple cooperation. I imagine the egg must be conspiring with the refrigerator door to piss me off.

"He's really in a foul mood this morning," whispers the egg to the refrigerator door, "so when he reaches for me, I'll stick to you, and you stick to me. Then when he tries to remove me, I'm sure to break. That will surely send him over the edge. What fun!" Why that conniving little egg. What perversity. "Is it that hard," I demand as though the egg could understand, "to just let go?"

OK, that mess is finally cleaned up. Guess I'll use some eggs that don't somehow have their twisted hearts set on dispersing goo over things. Once the eggs are cooking I decide to add some cheese, but it is wrapped in plastic – impregnably wrapped. Unassailable. Cheese packages no longer can be opened by hand. I am at a friend's house where the available knife doesn't cut it.

"What's wrong with all of you?" I demand. By now I am once again livid. The conspiracy has grown. It is not just eggs and refrigerator doors, but cheese manufacturers, plastic wrapper makers, and the plastic wrapper itself that are conniving to keep me from indulging in the breakfast of my choice. What to do? Ice pick? Screwdriver? Who or what can I stab? Where do they keep the scissors?

One thing, the egg, comes apart when I want it to stay together. The other, a cheese wrapper, stays together beyond reason when I want it to come apart. By natural law, these things happen only when we are at our most fragile.

How can we possibly deal with such perverted minds? Nothing – kicks, screams, depression, rage, pleading, wise explanation – seems to adequately communicate to them the necessity of simply *not* behaving that way. They don't seem to grasp the concept that everything could be much more amicable, if they just had more consideration for "me."

Once I stop to reflect, even a few minutes later, I realize that things don't conspire and come together with the intention of producing anger in me. Talk about stupid and misguided: How could I possibly attribute such identity and conspiratorial thinking to things? If I think so, I have a mistaken view about the way things behave. Anger feeds well on such mistaken understandings.

Who is it that needs to open up? Who is it that needs to let go? Guess I'd better get started. To dwell in the spirit of peace and harmony means realizing that there is no intelligence at work here plotting how to "get" me, and so no one at whom to get angry.

Breakfast, anyone?

Recipes

This breakfast menu is a combination of Patti's Omelet with my Corn Sesame Breakfast Cake. We both like prunes.

Patti's Omelet
Corn Sesame Breakfast Cake
Stewed Prunes with Orange and Cinnamon

Patti's Omelet

When Patti finally finishes her two hours or more of morning yoga, she'll sometimes make me this omelet, which she describes as the "sponge cake of omelets." There are several secrets to this dish which she has consented to reveal.

It all begins with eggs from free-range chickens with their deep golden yolks, rather than the watery, pale yellow-yolked eggs from chickens locked in wire boxes, unable to move, stuffed with pellets and antibiotics. The other keys are the yogurt, which makes for a fluffier omelet than milk, and having the pan medium hot initially and then turning it down low.

Serves 2 people, modestly

2 jumbo eggs, from free-range chickens

2 tablespoons plain yogurt

¼ cup freshly grated Asiago *or* Parmesan cheese

1 to 3 teaspoons butter

½ teaspoon fresh thyme, minced

Black pepper

Whisk the eggs in a bowl with the yogurt and freshly grated cheese. Let a 6-inch skillet warm up over medium heat. Add the butter – the lesser amount if the skillet is nonstick. Wait for the butter to stop bubbling, then pour in the egg mixture, cover with a lid, and reduce the heat to low. Check after about 3 minutes, and if the omelet is nearly firm on top, re-cover the pan, and turn off the burner, allowing the omelet to finish cooking from the heat of the pan.

Otherwise let the omelet cook a minute or two longer before turning off the heat.

Sprinkle on the fresh thyme and a light grinding of black pepper. Since Asiago cheese is salty, no additional salt is usually needed.

Corn Sesame Breakfast Cake

This is a somewhat unusual recipe: a cornbread that is more of a cake with a crust on the bottom. Don't know the real name for it, but it is adapted from the Turkish Coffee Cake Cookie Bars in The Tassajara Bread Book. *You probably have not had very many things quite like it. Enjoy.*

Makes 1 (8½ to 9½-inch) springform cake

1½ cups unbleached white flour

1 cup corn flour

¼ cup brown sugar

¼ cup white sugar

¾ cup butter

⅓ cup sesame butter or tahini

¼ cup honey

⅓ cup yogurt

1 egg

¾ teaspoon baking soda

Preheat oven to 375 degrees.

Combine the flours and sugars, and cut in the butter until it is in tiny lumps. Press about half of this mixture (about 2 cups' worth) into the bottom of an ungreased 9½-inch springform cake pan or an 8 x 8-inch baking pan. This will make the crust for the cake.

Combine the sesame butter, honey, yogurt, and egg. Mix the soda into the remaining flour mixture, then the combined liquids. Pour into the pan on top of the pressed-down crust.

Bake at 375 degrees for about 40 to 45 minutes until the center of the cake has risen and is bouncy to the touch. You can also stick in a toothpick and see if it comes out clean.

Stewed Prunes with Orange and Cinnamon

If we are cooking prunes we usually cook up extra, so that they can be reheated another day.

Serves 4 to 6 people

1 orange
1 pound prunes
3 cinnamon sticks

Cut the orange in half, then into slices (peel and all). Place in saucepot with prunes and cinnamon sticks, and add water to cover. Cook about 30 minutes until the prunes are quite tender, and the oranges are "melting."

Delicious!

Wishful Thinking Fails Again

Once during a cooking class I found myself holding a pan of biscotti in one hand, while I opened the door of the oven with the other. Since the top shelf of the oven was occupied by a pan of lasagna, I aimed the biscotti for the unoccupied bottom shelf. "Let's get these cookies baking," I thought. However, a second thought followed quickly after the first: "If you put them on the bottom shelf, they're going to burn."

Struggling to Get It Right and Eating the Results

What to do? There ensued a lively inner monologue enumerating the necessary steps: Close oven door, find somewhere to put biscotti down, open oven door, remove lasagna, find place to put lasagna, move top shelf up one notch, move bottom shelf up one notch, replace lasagna, close door, get biscotti, open door, place in oven, close door. What a nuisance. It hardly seemed worth it.

"Forget it," I thought. "Let's get on with it. They'll be okay this time." In they went.

These are a cook's famous last words: "This time they'll be okay." Sure. This time the oven will understand how awkward and inconvenient it is for me to do all that switching, placing, lifting, reaching, and it will go out of its way to accommodate me. The oven will make a special effort not to burn the cookies to compensate for my not making a special effort to arrange things differently. This time, undoubtedly, the oven will be forgiving and make allowances for my laziness. Only this time the cookies burnt on the bottom.

Once I used some vanilla sugar at a friend's house to make a birthday cake for my father. At least I thought it was vanilla sugar, since it was a white granular substance with a vanilla bean in it, and when I dipped my finger in it and licked, it tasted like sugar. Yet tasting the cake batter after creaming the sugar with the butter and adding the eggs, I found it extremely salty. And going back to the jar with the vanilla bean, it tasted like salt. Big surprise!

Not wanting to waste the butter and eggs, I decided to go ahead and finish making the cake, thinking rather wishfully that maybe it wouldn't be so bad once all the flour and milk and seasoning was added. It was. Yet my wishful thinking continued: "Maybe it won't be so bad after it's baked." It was. It was really bad, not at all what a cake should be. That was a strange birthday celebration.

The ability to believe in wishful thinking right up until you smell the smoke or taste the cake is really a wonderful trait in many ways – naive, trusting, childlike – but the food may be an uncustomary and undesirable shade of brown or black. The taste may bring tears to the eyes.

Although I still find it painfully annoying at times, the universe (including ovens and other cookware) does not arrange itself to pick up after me. Things are the way they are, regardless of how I would like them to be. If anything, it seems that the universe is conspiring to wise us up to our own wishful thinking. Would you wish it to be any other way?

Recipes

What follows is lasagna and biscotti, along with a salad. Just wishing won't be enough to make them appear in person to nourish your eyes, nose, tongue, and taste. Unless, of course, you give your wishing a hand.

Mushroom Ricotta (or Tofu) Lasagna
White Sauce
Five-Element Salad
Spicy Garlic Vinaigrette
Once-Baked Biscotti

Mushroom Ricotta (or Tofu) Lasagna

Preparing the lasagna will probably take a good hour and a half to two hours, but you will have a generous amount of flavorful, heart-warming food to show for it.

A friend's lasagna took even longer, so she served the appetizers and side dishes, and then invited her dinner guests back for lunch the next day. The main course was excellent. Alternatively the lasagna may be made up well in advance, then baked before serve-up. Also, it refrigerates and freezes well.

You will need about a pound of pasta. This recipe calls for generic lasagna noodles, but if you want to buy or make fresh pasta sheets, please do. The Greens Cookbook has two fine recipes. The fresh pasta does not need to be precooked, but if it is, your lasagna will attain even higher, more heavenly states.

Many people these days are not eating dairy. In that case, you can use more Herbed Tomato Sauce (see recipe page 246) instead of White Sauce (see recipe page 141), substitute drained, crumbled tofu for the ricotta, and perhaps some roasted, chopped walnuts or almonds for the Parmesan.

Serves 8 or more people

1 pound lasagna noodles

3 to 4 tablespoons olive oil

2 yellow onions, diced

3 stalks celery, finely diced

Salt

Black pepper

½ cup dry sherry *or* tomato paste, if not using alcohol

¾ pound mushrooms, sliced

3 to 4 cloves garlic, minced

⅓ cup parsley, chopped

1 teaspoon dried thyme *or* 1½ tablespoons fresh thyme, minced

1 pound ricotta cheese

2 teaspoons dried marjoram *or* 2 tablespoons fresh marjoram, minced

Peel of 1 lemon, chopped

2 teaspoons capers, chopped

¼ cup pitted olives, chopped

1 egg

White Sauce (see recipe page 141)

2 cups Parmesan *or* Asiago cheese (about 5 ounces), freshly grated

Herbed Tomato Sauce (see recipe page 246)

Flat-leaf parsley, for garnish

Heat a good 3 to 5 quarts of water to boiling, add a tablespoon of salt, and 2 or 3 of olive oil. Cook the noodles until they are tender, perhaps 7 to 10 minutes. Drain the noodles and rinse them in cold water to stop the cooking and to keep them from sticking together. Drain again, then spread them out individually on a clean dish towel so they will be flat when you are ready to use them.

Heat 1 to 2 tablespoons olive oil in a large skillet and cook the onion until it is translucent, about 3 to 4 minutes. Remove half of the onion, and set aside to add to the ricotta filling later. Add the celery to the cooking onions along with a sprinkling of salt and pepper. Sauté over high heat for another couple of minutes, then cover and cook over lower heat until the celery is tender. Remove from heat and set aside.

Use a quarter cup of the sherry to "deglaze" the pan, pour off, and save. Reheat the pan, add another 1 to 2 tablespoons of olive oil and sauté the mushrooms. After a minute or two add the garlic, parsley, thyme, and a touch of salt and pepper. Cook until the mushrooms are browned, then add both sherries (fresh and from deglazing). Let the sherry cook down by half, then add to the onion-celery mixture, and check the seasoning.

To make the ricotta filling, combine the ricotta with the reserved onion, as well as the marjoram, lemon peel, capers, and olives. Season with salt and pepper as needed, then mix in the egg. Set aside.

Prepare the White Sauce, below.

For the final assembly, preheat the oven to 375 degrees. Butter a 9 x 13-inch baking pan, and ladle a half cup of the White Sauce over the bottom. Cover the bottom with a layer of lasagna noodles. Coat the noodles with a few spoonfuls of the sauce, then add half the mushrooms and half the grated cheese. Next, put in a layer of noodles, a touch of the White Sauce, and half of the ricotta filling.

Repeat the layering: noodles, sauce, mushrooms, cheese; noodles, sauce, ricotta. Arrange a final layer of noodles on top, and pour over the last of the White Sauce, so that the whole lasagna (especially the last layer of noodles) is well moistened.

Cover with foil and bake for 30 minutes. Uncover and continue baking for another 10 minutes (as long as the top is still moist). More grated cheese could go on top.

While the lasagna is baking, prepare the Herbed Tomato Sauce (page 246), if you have not already done so.

Cut the lasagna into squares. For serving, ladle ⅓ to ½ cup of the Herbed Tomato Sauce onto individual plates, place a square of lasagna on top of the sauce, and garnish with several sprigs of parsley.

White Sauce (loosely speaking, béchamel)

3½ cups milk
4 tablespoons butter
4 tablespoons flour
Salt and black pepper
Nutmeg

Start the milk heating. Melt the butter in a saucepan, and stir in the flour to make a "roux." Cook for several minutes over moderate heat, taking care not to let the flour brown (if you want your sauce to be white). Remove from the heat and wait for the roux to stop bubbling, then pour in the hot milk and stir with a wire whisk.

Return to the heat and cook, stirring, until the sauce thickens. Season with salt and pepper, and a few scrapings of nutmeg, without adding so much that all you can taste is nutmeg. A tiny amount of nutmeg will bring up the flavors of the sauce. Remove from heat.

Five-Element Salad

I am not a fan of salads that have innumerable ingredients – lettuce, grated carrot, tomato, mushroom, bell pepper, shredded red cabbage, etc. – as though more made better, and then are drowned in some kind of gloppy dressing so the colors are indistinct anyway. I want the ingredients to fit together with some balance.

Five-Element is my name for a salad I learned about while working at Greens Restaurant. The five elements are lettuce, a fruit, a nut, a cheese, and a wild-card flavor kicker, such as olive, caper, radish, sun-dried tomato, or, in this case, red onion pickle. The dressing should be flavorful and interesting, but usually clear enough to see the colors of the salad. The variations are endless.

At home I am more likely to prepare a Garden Salad (page 188) but occasion-

ally I am more inspired, especially if company is coming. I follow the Chinese principle of having an odd number of ingredients: one, three, five, seven. See how that works for you.

Serves 4 to 6 people

1 head romaine lettuce, moderate sized, about ¾ pound

⅓ cup almonds

1 apple, a good eating variety such as Fuji, Gala, or Golden Delicious

Juice of 1 orange, preferably navel

Spicy Garlic Vinaigrette (see next page)

½ cup Red Onion Pickle (see recipe page 17)

½ cup Asaigo *or* Parmesan cheese, grated

1 tablespoon fresh tarragon, minced, *or* 2 tablespoons flat-leaf parsley, minced

Remove the bruised, discolored, or wilted outer leaves from the head of lettuce. Cut the larger romaine leaves in half lengthwise and then crosswise into 1-inch sections. Cut the smaller leaves crosswise into 1-inch sections. Wash and spin dry.

Roast the almonds in a 350-degree oven for 8 minutes (or in a dry skillet on top of the stove) until they are crunchy and aromatic. Slice or chop them.

Quarter the apple, remove the cores, and cut into slices. Combine with the orange juice.

Prepare the vinaigrette.

All the ingredients may be readied well ahead of time and assembled shortly before serving. The point to notice is that if you toss all the ingredients together, the smaller ones will end up on the bottom of the bowl. Here's what to do: Start by combining the apple slices (orange juice and all) with the lettuce and tossing with the vinaigrette. Make sure some apple slices are at the surface. Distribute the red onion pickle over the surface of the salad, then the almonds, the grated cheese, and finally the minced herbs.

Spicy Garlic Vinaigrette

This could be called a curry vinaigrette, since the same spices are often used in Indian cuisine.

- 1 teaspoon fennel seed
- 1 teaspoon coriander seed
- ½ teaspoon cardamom seed
- ¼ cup olive oil
- 3 tablespoons balsamic vinegar
- 1 clove garlic, average size, minced
- ½ teaspoon salt
- ¼ teaspoon black pepper

Combine the fennel, coriander, and cardamom seeds and grind in a coffee mill or spice grinder. Whisk together with the olive oil, vinegar, garlic, salt, and pepper.

Once-Baked Biscotti

Bis is the word for "again," indicating that these cookies are twice-baked, but I much prefer them once-baked. Perhaps we should call them uns-cotti. Even once-baked they keep well, and are quite pleasant with tea or coffee. This recipe uses anise seed and fresh orange peel in place of orange and anise extracts to make the flavors brighter and fresher.

Makes 4 dozen or more

- 1 cup unsalted (sweet) butter, softened
- 2 cups white sugar
- 2 eggs
- 2 tablespoons anise seed *or* 2 teaspoons anise extract

Peel of 2 oranges *or* 1 tablespoon orange extract

3 cups unbleached white flour

1 cup whole-wheat flour

2 teaspoons baking powder

Preheat the oven to 350 degrees.

Cream the butter and sugar together, then beat in the eggs.

Mince the anise seed with a knife, or blend in a spice mill.

Remove the orange peel with a vegetable peeler, then chop it or cut it into thin strips. Add the seeds and peel to the creamed mixture. Mix the flours together with the baking powder, and combine well with the creamed mixture.

Divide the dough into 4 pieces and shape each one into a log about 12 to 15 inches long. Put 2 on each baking sheet, and bake until the bottoms are browned and the top surfaces are cracked, about 20 to 25 minutes. The "logs" will have flattened out sideways. Let them cool a few minutes and then cut crosswise into ¼-inch pieces.

If you insist on a second baking, place the biscotti on their sides in a 300-degree oven until they dry out, about 10 to 15 minutes.

Once the biscotti are completely cooled, they may be stored in a tin.

Appreciating the Pure Flavor of Kind Mind, Joyful Mind, and Big Mind

The Roshi Moves a Rock

As most of us know from first-hand experience, the gift of food nourishes our spirit as well as our body. We feel that Divine Providence or someone in particular cares about us, and we may offer thanks or gratitude in return. A word or gesture can uplift or cast down, and a material gift can have the power, as does food, to nourish far beyond the substance itself. Something of the heart is conveyed and received.

From May 1967, when I returned to Tassajara to be the head cook, I lived in the same room for nearly three years. It was in the back of the first cabin across the bridge heading toward the swimming pool and had its own toilet and sink with cold running water. Those who lived in the front room used my room as a passageway to and from this bathroom.

Sitting in this same room today while I write more than twenty-five years later, I find it much more pleasant because of the addition of a large picture window on the back side overlooking Tassajara Creek. I can gaze at the bay and alder trees and the occasional maple branches showing here and there. On the far side of the creek a steep hillside is home to a number of oaks. It is a sweet view that I didn't get to enjoy when I was living here previously, but since that time my teachers found various ways to make "improvements" in the space.

In the Japanese Soto Zen tradition the cook is encouraged to cultivate the three minds: Joyful Mind, Kind Mind, and Big Mind. So Kobun Chino, one of our teachers, wrote out the Japanese characters on three small plaques made of cross sections of a tree branch. These were attached to the door of my room so that I saw them whenever I returned.

Coming back to my room I would be reminded that someone was thinking of me, someone believed in me. And I would be reminded of the mind which is buoyant and rejoices in offering its efforts for the benefit of others; the mind which nurtures and cares for all beings and things as a parent does for a child; and the spacious mind which attends equally to each thing small or large, and where everything has its place.

I certainly couldn't "maintain" those three minds, but sometimes I would get a taste. One time Suzuki Roshi gave me a taste of his mind. I was standing on the bridge with him, and he remarked that the pile of rocks outside my door, which I was using for steps, was rather unsightly. "In Japan," he continued, "sometimes we pile rocks like that on a grave. Those rocks have a forlorn and melancholy feeling; feels not so good, you know."

I explained that I didn't know much about working with rocks, so I had just piled them up the best I could. "Yes, you're right," I said. "They don't look so great, and you know what else – they wobble when I step on them." And I tipped my hand back and forth in the air to see if I could make him grin. He nodded and went back to viewing the rocks and the creek, serene and impassive. You never knew what he was thinking.

Weeks later when I had forgotten all about the rocks, he stopped me as I was leaving his cabin, as though he had just remembered some small matter. "Oh Ed," he called out, waiting for me to turn back to the room, "do you know that rock in front of the office? I asked Paul to move it to your cabin this afternoon to be your doorstep. Is that OK?"

I was flabbergasted. The big flat rock outside the office had a central place in the community. Often there seemed to be a cluster of people sitting around it and reading their mail or smoking cigarettes. "That rock," I answered, "is a real focal point for the community." "We'll get another one," was his response.

Sure enough, that afternoon as I sat in my cabin, I began hearing the sound of the rock being towed along the ground in a metal sled behind the power wagon. Coming across the bridge, it produced an especially loud screeching, and then the sound stopped as the truck halted at the front of my cabin.

I went outside, and there were Paul and Roshi and one or two others, and I think they used two-by-fours for track and short pieces of two-inch pipe for rollers to move the rock from the roadway to my door. The pile of rocks that had been my steps was heaved over the edge into the creek bed. Then they moved the new rock into place. About five feet long and a foot and a half high, it was a big beautiful, flat, solid rock – grey with streaks of tan and beige, an occasional vein of white, somewhat pointed at one end and rounded at the other.

Everyone seemed quite happy, and I couldn't believe how wonderful and reassuring it felt to step on a big, stable rock instead of that uneasy pile of stones. Nothing wobbled, and I felt a tingling and a warm joy as something more solid settled into place inside me. In and out of my cabin, stepping up and stepping down, my teacher was there for me. My teacher supported me, gazed back at me, accepted me. These are the best gifts, the ones that move us to feel how very deeply we can trust the universe, trust our own body and mind.

When the spirit is fed, we grow larger-hearted, and we want to pass on the gift.

Recipes

This menu is one I find quite reassuring somehow, like a gift. The two dishes work well together and are easy to prepare, yet the flavors are robust and remind me of the sustaining capacity of earth, air, sun, and water.

Potato Gratin with Celery Root and Fennel
Chard with Lemon and Raisins

Potato Gratin with Celery Root and Fennel

I like the word gratin *here better than* casserole. *Doesn't it sound better to you?*

Serves 4 to 6 people

3 to 4 red potatoes (about 1½ to 2 pounds)

1 small celery root (about 1½ pounds)

1 large fennel bulb or 2 smaller ones (about ½ pounds)

6 to 8 cloves garlic

Olive oil

1 tablespoon fresh thyme, minced, *or* 1 tsp dried thyme

Salt

Pepper

Fresh thyme leaves, minced, for garnish, if available

Cut the potatoes into eighths or smaller. Rinse any excess dirt off of the celery root, and then trim off the skin and rootlets all the way around. Cut into quarters and then crosswise into slices an ⅛-inch or so wide. Cut the stems off the fennel (sweet anise) bulb, but save the soft feathery "leaves." Cut the bulb in half, then crosswise into slices. Mince the "leaves." Peel the garlic cloves and slice into halves or quarters.

Oil a baking dish or pan, and put in the cut vegetables along with the thyme, salt, and pepper. Cover and bake in a 375-degree oven for 75 to 90 minutes until the vegetables are very tender. Wonderful. Garnish with the minced green of the fennel and some fresh thyme, if available.

Chard with Lemon and Raisins

Serves 4 to 6 people

1 bunch chard

⅛ lemon

1 leek or medium yellow onion, diced

¼ cup pine nuts

1 tablespoon olive oil

⅓ cup raisins

Salt

Pepper

Wash the chard and cut the stems crosswise into narrow pieces. Cut the leaves in half, then crosswise into 1-inch pieces. Cut the wedge of lemon crosswise as thinly as you can, peel and all. Slice the white part of the leek and wash the slices to remove any dirt beneath the surface.

Toast the pine nuts in the oven or on top of the stove until browned (see page 13). Be careful not to burn them.

Sauté the leek slices in the olive oil for 2 to 3 minutes. Reduce the heat, and add the chard, lemon, and raisins along with a sprinkling of salt and pepper. Cover and cook until tender. Garnish with the roasted pine nuts.

The Workers Rebel, the Real Work Deepens

After I had been the head cook at Tassajara for a year or so, a kitchen rebellion broke out. At first I couldn't understand why. I was sincere and responsible, I worked hard, and as far as I could tell, I meant well. Yet the crew had decided I was dictatorial, autocratic, and unfeeling. They had gone to the heads of the community and complained about me.

At a meeting to air their grievances one woman said that I treated everyone as though they were just another utensil in my hand, that I didn't acknowledge that they also could make decisions involving taste. "It's not as though you are the only one who can cook, you know, but you never let us. You always make all the decisions." She wanted me to understand that there was a pivotal difference between herself and spatulas – she could think, feel, see, taste, and make aesthetic decisions.

Another woman said, "You treat us the same way you treat the bread," and then backtracked briefly, by saying, "Actually you treat the bread pretty well. You treat us *worse* than you treat the bread." Couldn't I be as sensitive to her and her needs as I was with the bread dough?

Later the director of the monastery, Peter Schneider, came to talk with me. "You need to give people more responsibility," he said. "Would you be willing to change the way you do things, or would you like another job?" I felt devastated and humiliated. I couldn't imagine doing things differently. "Think it over," he concluded, "and let me know."

I sat outside in the sun and cried, feeling lost and disoriented. I had done the best I knew how to do, and that was being trashed. I was shocked to learn that people saw me as using them.

Trudy Dixon came over and sat down next to me. Trudy was a senior member of the community whom I respected, and for her to take an interest in me was a surprise, especially since she was fighting cancer and probably would not live much longer. Yet she took the time to listen to my stumbling account of the situation, as well as my bewildered "I just don't know what to do."

Her response, "I believe in you," stunned me. I couldn't believe it. By now the sun was even more warming, and I was touched that this person who had worked on *Zen Mind, Beginner's Mind* with Suzuki Roshi would say such a thing. Still, I protested that her faith was probably misplaced, and she simply repeated herself: "I have faith in you." It made a world of difference.

So I did take some time to think over what I had been doing. Why was I working so hard, and not really letting anyone help? When I considered allowing others to do more of the "cooking," I realized I had been making a tremendous effort to impress and astound people with my cooking artistry. If I let others do more of the cooking, it wouldn't be "mine" anymore, and I would not have the same supposed fame in which to bask. And I wanted to be great. Then people would like me. Then women would love me.

And what difference was that going to make? If I could get others to love me, maybe that would convince me to love myself. Only by this circuitous route was I coming to notice that I didn't like myself. What a surprise that I hadn't realized it sooner. And then I saw how hard it was to please me. Yes, I would be willing to love the most perfected being ever.

Thus was I caught in my own mistaken effort. Attempting to become perfect enough to be lovable was clearly a useless endeavor. Besides, if my self-esteem depended on accomplishment or performance, then I would only be "as good as my last meal" and always have to keep surpassing myself in order to earn my own and other people's love. My self-esteem would be inherently fragile.

Besides, who could say for sure that if someone liked my cooking that meant they liked me? Quite probably they didn't really know or care about me at all and just wanted my cooking to continue unabated. To paraphrase a Buddhist teaching, "I am not my cooking. My cooking is not me."

If I wanted to like myself, I'd have to go about it more directly, and have more compassion for someone rather ordinary with problems: me. That was my work and not the work of the food. Let the food be food, and speak for itself.

Still, if I wasn't going to cook to "prove" anything, why bother? Raising this question brought into focus other threads of motivation: basic kindness and generosity, the wish for the happiness of all beings. I would cook because I wanted to cook. I would cook because I wanted to offer food, so I could *be* food.

Recipes

Time to offer you a menu and see what you think. Any changes? Simplifications? Elaborations? I encourage you to improve upon anything I might suggest.

Corn Timbale with Ancho Chili Sauce
Broccoli with Olives and Lemon

Corn Timbale with Ancho Chili Sauce

The aim of this timbale is to have a pure, strong, singular taste of corn to balance the intense flavors in the Ancho Chili Sauce.

Timbales are egg custards that are customarily made with pureed vegetables. They can be excellent vegetarian entrees, and there are a number of good ones in The Greens Cookbook, *which are often made with milk and grated cheese, in addition to the eggs and vegetables. In this case the idea is* corn, *basically corn and water. The smooth texture and clean flavor make the dish soothing and honest.*

Makes 4 larger or 6 smaller portions

3 ears of corn

4½ cups water

1 small yellow onion (about 6 ounces), diced

2 tablespoons butter

Salt

Black pepper

3 extra-large or jumbo eggs

Ancho Chili Sauce (see recipe page 243) or one of the other tomato chili sauces
    (see recipes pages 244–45)

Fresh cilantro sprigs or leaves, for garnish

Preheat oven to 325 degrees. Have some extra water boiling on the stove. You will need it when you are ready to bake.

Shuck the corn and carefully remove all the silk. Laying the corn down flat on the counter, cut along the edge of the cob to cut off the kernels. Rotate the ear of corn and cut off another strip of kernels. Continue until the kernels have been cut off all the way around. Set the kernels aside, and start the cobs cooking in the 4½ cups of water. Bring to a boil, then reduce the heat so that the "stock" simmers for 30 to 40 minutes. Once the stock is "finished," remove the cobs.

Sauté the diced onion in 1 tablespoon of the butter in a large skillet for a few minutes, then add the corn kernels and continue cooking over moderately high heat. Once the kernels have been roasted a bit, add ¼ cup of the stock, sprinkle with salt and pepper, cover, and reduce the heat. Cook over low heat for 15 minutes or so, until the corn is quite tender.

Put the corn and onion in a Cuisinart or blender with 2 cups of the stock, and puree several minutes until the mixture is quite smooth. (My immersion blender does not work as well for this.) Check the seasoning.

Whisk the eggs in a bowl and then slowly whisk in the pureed corn.

Butter 4 or more ceramic ramekins or ¾- to 1-cup (small) ceramic bowls. These need to fit into a sided baking pan – a 9 x 13 x 2-inch one is probably about right. Fill the dishes with the timbale mixture to within ¼-inch of the tops. Place in the baking pan, and place the pan in the oven.

Pour the extra boiling water into the baking pan around the timbale dishes. The timbales will bake much more quickly if boiling water (rather than tap water) is added at this point, even if you have to take a few minutes to boil it. And pouring the boiling water once the pan is in the oven obviates the necessity of lifting and moving a pan of boiling water and timbale dishes around the kitchen, and getting it into the oven without spilling. Bake at 325 degrees for 50 to 70 minutes until the tops puff up and are firm to the touch.

While the timbales are baking, make the Ancho Chili Sauce, if you have not already done so.

Remove timbales from the oven and let cool 5 minutes. Ladle the sauce onto individual plates for serve-up. Then slip a rubber spatula around the sides of each timbale, and turn it out upside down on top of the sauce.

Garnish with the cilantro sprigs or leaves.

Broccoli with Olives and Lemon

The olives and lemon give this dish something of a Mediterranean feeling.

Serves 4 people

1 bunch broccoli

1 tablespoon olive oil

1 red onion, sliced

3 cloves garlic, minced

1 teaspoon sugar

Juice of ½ lemon

Peel of 1 lemon, minced

Salt

Black pepper

12 Kalamata olives (about 1½ to 2 ounces), pitted and minced

Top the broccoli and cut it into fork-size pieces. Cut the tough skin off the stalk and then cut the stalk into fork-sized chunks. Many people consider this tender inner portion of the stalk to be the best part. Give it a try.

Heat the oil in a skillet and sauté the red onion for 2 to 3 minutes. Then add the garlic and the broccoli along with the sugar, and continue sautéing for another minute or two. Add the lemon juice and peel, and if the vegetables begin to stick, a small amount of water. Sprinkle on some salt and black pepper, cover, and cook over moderately low heat until tender, about 6 to 10 minutes. I like the broccoli when it is beginning to get tender and is still bright green. Stir in the olives and continue cooking just until heated. Check the seasoning. If you find the lemon too strong, you can always use less, or add a bit more sugar.

Empowering People to Cook

People living together in a meditation community are like rocks in a tumbler. Spinning around and around, constantly bumping into one another, the rocks get worn down, smoothed, polished. Sometimes we would joke about it: "Some say polished, some say ground." The process itself is not always a pleasant one, but there was no escaping it, no matter how hard we tried. Family life, life in the world, can also polish us.

As the cook at Tassajara I started out thinking that creating delicious food was the most important thing. So I composed the menus, and I decided who did what, and I finished the seasoning on each and every dish. I did things my way, the way they should be done. I was a young, inexperienced first-time manager, no doubt about it: Everybody should work the way I work, do things the way I do them. Wasn't that what being in charge was all about?

After the kitchen rebellion I felt embarrassed about going back to work there – vulnerable and exposed. Where could I ever be safe? The only safety was to trust in others, to trust in the life we were living together, trust in being revealed. Some say polished; some say ground.

The more deeply I looked, the more I had to admit the limitations of my efforts. Others had been deferring to me and not really taking responsibility for what happened in the kitchen, because I didn't give it to them. I always had to be there, since others could not function without me. Providing the answers meant I was important. And being indispensable meant I kept pointing out how great I was, how retarded they were. No one could grow. No one really got to develop his or her capacities.

How strangely disillusioning – that being important and indispensable meant being stuck. Others had been just as snared by the dynamic, yet they had postponed confronting me directly. When they did, not surprisingly, they were angry. After all, they had agreed to making me look good, while disowning their own capabilities.

When I went back to work in the kitchen, I began acknowledging other people's abilities, and letting them make decisions of consequence. I started taking regular days off and turning the kitchen over to one of the other members of the crew, each in turn. I felt an unfamiliar tenderness or compassion, since I knew that despite our best efforts, we can be so easily belittled or dismissed.

Of course as more members of the crew experienced the responsibility of generating meals, they became more sympathetic to what I had been going through. We began to share something: the work of the kitchen, the space of the kitchen, some comraderie. I stopped feeling so alone and isolated.

Gradually I began to see what I was "doing" in a new way. Instead of striving so desperately to be indispensable, what I needed to do was train people to cook, train a successor, make myself dispensable.

Encouraging responsibility is often a delicate matter. When I worked at Greens Restaurant many years later I saw how easily responsibility can be given and taken back. The expeditor, who garnished and assembled each order, also had the responsibility of seeing that the next day's prep was done. But the head lunch cook would say, "You've got to do the potatoes now." So who's in charge? Who has responsibility? Not letting someone fail takes back the responsibility. To have responsibility means experiencing the consequences of one's actions.

Accountability, I found, was different from responsibility. If the head cook had inquired, "How's the prep going? What's left to be done?" then the expeditor would still have had the responsibility, but would have been obliged to give an accounting of how it was going and would probably have realized what remained to be done.

As a cook I had focused on the mechanics of kitchen work: cutting and scrubbing, organization and menu planning, inventory and ordering. Now I was finding out that empowering people to cook was much more gratifying. Training cooks, finding a successor – that was a goal worth working for. It has turned out to be a wonderfully engaging activity which I have spent years working on, decades, really, and it all came out of people's unhappiness with me. Out of it came cookbooks and managing Greens, cooking classes, and workshops. And I found that when I empowered people to cook, the food would take care of itself.

May you also awaken in others the thought of baking bread, washing rice, stirring soup, cutting carrots . . . endlessly.

Recipes

Of course, empowering yourself to cook is just as important as empowering others. This means giving yourself permission to try things out, to find things out. You allow your curiosity and enthusiasm some space to cavort, tasting and experimenting with unfamiliar (and familiar) ingredients. Although you may have a tentative plan or agenda, still you leave yourself space to "follow your nose" and "play it by ear."

Here is a menu that is more elaborate than most of the ones I am sharing with you – one that could graciously entertain guests or friends – yet I think you are ready for it. I support you in your efforts.

Winter Squash Soup with Apple, Cumin, and Cardamom
Mushroom Filo Pastry with Spinach and Goat Cheese
Beet Salad with Watercress

Winter Squash Soup with Apple, Cumin, and Cardamom

Serves 4 to 6 people

1 or 2 winter squash, such as pumpkin, Perfection, acorn, Delicato (about 2 pounds)
1 teaspoon cumin seeds

¼ teaspoon cardamom seeds

1 yellow onion, sliced

1 tablespoon olive oil

2 cloves garlic, minced

1 tablespoon fresh ginger (about 1 ounce), grated

1 apple, cored and sliced

4 cups hot water

1 tablespoon lemon juice

Salt

Bake the winter squash about 1 hour at 375 degrees. Allow it to cool, then cut open, remove seeds, and scoop out the flesh.

Grind the cumin and cardamom in an electric coffee mill used for grinding spices. Sauté the onion in olive oil for 2 to 3 minutes, then add the garlic, ginger, cumin, and cardamom and continue cooking another 1 to 2 minutes.

Add the apple and 4 cups hot water, along with the squash. Cook for 10 minutes or so. When the apple is soft, puree with a hand blender or Cuisinart. Season with the lemon juice and salt to taste.

Mushroom Filo Pastry with Spinach and Goat Cheese

Filo pastries are wonderfully festive, and even given the amount of work involved, they go a long way, so I have made this dish for several benefit meals.

Serves 6 to 8 people

½ pound filo pastry

¾ cup pine nuts

1 large or 2 smaller leeks, white and pale green parts

½ pound spinach

½ pound mushrooms, perhaps fresh shiitake or portobello as well as cultivated
      mushrooms

1 tablespoon butter

3 cloves garlic, minced

1 tablespoon olive oil

1 tablespoon fresh thyme, minced *or* 1 teaspoon dried

1 teaspoon fresh rosemary, minced *or* ¼ teaspoon dried

1 tablespoon fresh marjoram, minced *or* 1 teaspoon dried

Salt

Black pepper

¼ cup white wine

4 to 5 ounces goat cheese (chèvre), crumbled

½ cup ricotta cheese

2 eggs

4 tablespoons unsalted butter

4 tablespoons olive oil

24 whole cloves

12 sprigs of flat-leaf parsley *or* fresh thyme, rosemary, or marjoram

Preheat oven to 350 degrees.

Thaw the filo pastry (which usually comes frozen and folded up). Remove ½ pound by unfolding the pastry sheets and cutting them in half. Refold and repackage the unused portion and refreeze or refrigerate. (While the filo is thawing, you can be working on the rest of the recipe.)

Roast the pine nuts for 6 minutes in a 350-degree oven. Then chop them coarsely. Raise the oven temperature to 375 degrees.

Slice the leeks, then wash them by immersing in water and drain. Cut off the large stems of the spinach and cut the leaves crosswise into 1-inch lengths. Wash the leaves with plenty of water to remove all the sand and silt, and drain or spin dry. Slice the mushrooms, removing and discarding the stems of the shiitake or portobello mushrooms.

Heat a large skillet, add the tablespoon of butter, and sauté the mushrooms. After a couple of minutes add half of the minced garlic and continue cooking until the mushrooms brown. Remove from the pan and set aside.

Clean out the pan, reheat, add the tablespoon of olive oil and sauté the leeks for 3 to 4 minutes. Add the other half of the garlic, the thyme, rosemary, and marjoram, and a sprinkling of salt and pepper. Cook another minute or two. Add the white wine and continue cooking until the wine reduces by half. Return the mushrooms, add the spinach, and cook, covered, until the spinach wilts. Check the seasoning.

Remove the vegetable mixture to a bowl and add the goat cheese, ricotta, and eggs.

Melt the unsalted butter with the olive oil.

Assemble the filo pastry in a 9 x 13 x 2-inch baking pan. Trim the sheets of filo pastry so that they will fit in the pan: The sides of the pan angle apart, so make the sheets closer to the size of the top of the pan rather than the bottom. Those at the bottom can come up the sides a little.

Layer 7 to 8 sheets of filo pastry into the pan, one at a time, brushing each layer with the butter and olive oil mixture, and sprinkling with a scattering of pine nuts. Spread the filling on top of the upper layer. Top the filling with the remaining layers of filo pastry, brushing each with the butter/oil and sprinkling each with pine nuts (until they are gone).

Using a sharp knife, cut the filo pastry into 12 squares. Then cut the squares in half diagonally. Place a whole clove in the middle of each triangle. Bake in a 375-degree oven until the top is well-browned, about 40 or perhaps 50 minutes.

Serve 1 or 2 squares per serving, garnished with a sprig or two of fresh herbs.

Beet Salad with Watercress

Serves 4 to 6 people

5 to 6 large or 10 to 12 smaller beets (about 1 pound)
1 bunch watercress
2 green onions

¼ cup balsamic vinegar

3 tablespoons honey

¾ teaspoon allspice

Salt

Black pepper

1 tablespoon olive oil

Baking the beets rather than boiling or steaming will give them a richer flavor. Leave an inch of stems on the beets and place them whole in a baking dish. Add about ½ inch of water, cover, and bake in a 375-degree oven for 45 to 60 minutes. Remove and let cool, then slip off the stems and skin by hand. Slice the beets.

Remove any tough stems, then wash and dry the watercress. Thinly slice the green onions. Save some of the green for garnish, then mince the rest of the onions, especially the white part.

Combine the vinegar and honey. Use about ⅔ of it to dress the beets, along with the minced green onion. Then season the beets with the allspice, and some salt and pepper.

Toss the cress with the olive oil, then with the remaining vinegar-honey mixture and some salt and pepper. Serve the beets and cress side by side and garnish the beets with the remaining green onion slices.

The Sincerity of Battered Teapots

In the late sixties, when I was working so very hard and struggling to learn how to cook and how to direct the operations of a kitchen, the battered teapots were one of the things that kept me going. Dented and tarnished they sat on a shelf in the kitchen, ready to be used when called upon. My tired, despairing eyes would wander around the kitchen at all the jars and bowls, pots and utensils which were so much a part of my busy life and finally come to rest on those teapots. How did they do it?

Once they had been new, bright, perfect, a softly lustrous golden tone. Made of polished metal, probably aluminum, they were pleasingly round and plump, with a long perky spout and a graceful curving metal handle wrapped with bamboo stripping.

The teapots were used several times a day to serve hot water and tea. To see them filled and waiting was a cheery sight, not just because of the hot refreshing liquid stored inside, but because their shape greeted the eye with easygoing ampleness. Nothing pretentious, sleek, or stylish distinguished these teapots, which were always ready and always willing.

Zen offers a simple dictum for how to care for things, how to respect them: Carry one thing with two hands, rather than two things with one hand. The teapots rarely received this respect. Especially once they were empty, people would grab two handles in one hand and two handles in the other, and the teapots would clang their way back to the kitchen.

To practice respect or to care for something or someone intimately takes time, and even spiritually minded zen students are as much in a hurry as the next person. Instead of dashing to work or school, the zen student races for time off, a nap, or a hot bath. Teapots become an obstacle between here and rest, so grabbing two pots in each hand seems like a great time-saver.

After a while the teapots reflected the way they had been treated. Gazing at the teapots on the shelf, I would feel a certain comraderie: I too am like that – dented, discolored, drained. Yet as I looked I would sense something else: quiet dignity . . . tremendous forgiveness . . . the willingness to go on. "Sweethearts," I would think, "if you can do it, I can too." Inspiration comes from the strangest places.

Please do not suppose that I am condoning abuse. It is just that we all get beaten down by life – with disappointments and frustrations, annoyances and fatigue. And somehow we find the strength to continue. And sometimes the courage to change.

Recently my sweetheart Patti asked me if I knew what *sincere* meant. She had been working endless hours on a figure sculpture, which was to be cast in bronze eventually.

"No," I replied, "tell me about the meaning of *sincere*." Her explanation was that the *s-i-n* was like *sans* in French, meaning "without," and that the *c-e-r-e* meant "wax." To be sincere is to be without wax, the wax which can be used to cover up all the dents and blemishes, the chips and cracks, all those places we think we need to hide.

To be sincere is to be of a piece – with the imperfections showing. The lines and grooves are part of the beauty. The faults and shortcomings are part of the sincerity. When it comes to cooking, I put my faith in sincere, honest effort. I am less interested in showy, dramatic results intended to impress and astound than in day in and day out cooking. According to an old Chinese saying, "The uses of cleverness are soon exhausted, while the apparently simple is infinitely interesting."

To be committed to covering up faults is to be continuously anxious that we could be unmasked or seen through. When the imperfections are pointed out, we can become angry or quite discouraged. Yet although we are "up-set," this removal of wax can also be a relief. Then we don't have to put all that effort into covering up anymore. The secret is out.

In one zen story the student asks the teacher, "How can I attain liberation?" and the teacher responds, "Who is binding you?" The student is said to have had an awakening.

I find that story revealing. I notice how I bind myself at times with demands for perfection and mastery. I tell myself endlessly, "Watch what you say. Watch what you do," until a kind of paralysis sets in. I withhold love and respect from someone who is dented and tarnished, and even find fault with his efforts to wax things over.

Then I look at the teapots. And I am released.

Recipes

Speaking of sincere and full of heart, I think of barley mushroom soup. After a few too many of those rich, pretentious dishes or those fast-food delicacies, this is a soup that can help you get your feet on the ground or, as Kabir says, "Enter into your own body. There you'll find a solid place to put your feet." Can you stand it? The Chinese cabbage salad is a cool, juicy, refreshing accompaniment.

Barley Mushroom Soup
Chinese Cabbage Salad with Oranges and Mint

Barley Mushroom Soup

Wonderfully down-to-earth, this soup is quite simple to prepare. Mostly it just takes time.

Serves 4 to 6 people

½ cup barley

5 cups water

1 small red onion (about 4 ounces), diced

2 stalks celery, diced

1 carrot, diced

8 to 10 mushrooms or more (about ½ pound), sliced

¼ teaspoon salt

1 teaspoon dried thyme

Black pepper

⅓ cup flat-leaf parsley, minced

Cook the barley in the water over moderate heat for 30 to 45 minutes, so that it is well-softened. Add the red onion, celery, carrot, mushrooms, salt, thyme, and some black pepper. Cook an additional 30 minutes for the vegetables to soften. Add more water if needed. Mix in a generous amount of parsley to finish, and check the seasoning before serving.

Chinese Cabbage Salad with Oranges and Mint

For a while I hesitated to emphasize how good Chinese cabbage is in salads, thinking that it may not be that widely available, but then I discovered it in St. Johnsbury, Vermont, and Halifax, Nova Scotia. I make this salad quite plain, but you are welcome to season it as you see fit.

Serves 3 to 4 people

½ small head Chinese cabbage (also known as napa cabbage)
Salt
3 navel oranges
2 tablespoons fresh mint, chopped

Cut the half cabbage in half and cut out the core. Cut crosswise on the diagonal into fairly thin slices. Put into a bowl, and sprinkle on a bit of salt, perhaps ¼ teaspoon. Work it into the cabbage, gently squeezing with your hands.

Cut the peel off the oranges and then cut the oranges into rounds (see page 67).

Mix together the cabbage, oranges, and mint.

Note: If you do not have fresh mint available, use some minced parsley and a few pinches of ground fennel or anise seed.

Finding Out That Food Is Precious

Food is precious. We don't always remember until not much is left. Then it is obvious. Food is an everyday matter, until it disappears. Then we know it's terribly important. The simplest dishes can be divine. The fact that they are even there is providential.

The winter of 1969 at Tassajara we found this out. We hadn't stocked up as well as we might have with grains, beans, seeds, nuts, dried fruits, and canned goods, so when the rains came we were ill-prepared. And the rains did indeed come – torrentially. Tassajara Creek, which is placid in the summertime, swelled so that it roared through the canyon.

At the baths the stream rose a good ten feet or more and threatened to flood into the bathhouse. Perhaps that was the year the bridge to the baths was swept away, and someone climbed a tree that arched across the creek and dropped down to the baths. (We later installed a rope bridge.) It's all rather fun and exciting, grand and dramatic, as long as one is dry and has enough to eat. When one starts going hungry, then the constant, pulsating, throbbing surge of water is no longer thrilling, but takes on ominous qualities. What are we up against?

News came back that the fourteen-mile dirt road to town was no longer passable. About two miles up the road from Tassajara a huge mud slide had deposited a rock the size of a small cabin on the road, and nearby runoff had washed out a tremendous gully about six feet deep directly across the road. Not that we were accustomed to going out to the movies anyway, but we did depend on the road for food supplies. Further up the ridge, the road was impassable with snow and fallen trees.

We became foragers. Four people went out each day to pick the wild miner's lettuce and curly dock, which were plentiful because of the precipitation. At lunch, after the wheat flour for bread was gone, I began to serve white rice watered down to make a kind of gruel or congee. With it we would have a soup made with what we could find – perhaps a few beans or yams – and a miner's lettuce salad. Dinner would be brown rice with some steamed curly dock.

Once or twice we managed to get food in. After the county road crew had cleared the snow and fallen trees, a truck would come as far as it could, and we would drive up the road as far as we could, and then "ferry" food by hand or wheelbarrow over and around the various obstacles. Then the road became impassable again.

The food staples were stored in a room overlooking the creek, and sometimes I would look from one to the other: A few meager bags and tins of food staples, a magnificent surging of water. "This," I would think, "these few bags, are all there is. We could actually run out of food." Would we have to abandon Tassajara and hike out? We did have a large sack of whole-wheat berries left over from our failed efforts to grind our own flour, so I began cooking them and adding them to the rice gruel at lunch. With spring we began adding lavender-colored lupine blossoms to our miner's lettuce salad and sprouting some alfalfa seeds that we had discovered.

When the road remained impassable, we arranged for a helicopter to deliver food to Church Creek Ranch, which was about a three-hour hike, eight miles away. We trekked over and then back, our packs laden with groceries, some of which we wouldn't have ordered, but ate anyway: Jiffy peanut butter, Mary Ellen jam, Wonder bread. We had these fantastic giant gems called oranges. We had cauliflower and broccoli, cheese and nuts, onions and garlic; we feasted.

Not so long after that the road crew, which had been waiting for dry weather to stabilize the hillside, came in and cleared the road, dynamiting the giant boulder that had blocked the way. We'd survived an unusual winter.

Later that spring the toilet in the dormitory backed up, and Reb, our plumber (who later became one of our abbots), was asked to investigate. Opening up the drain line, he discovered that it was blocked with undigested intact whole-wheat berries. Live and learn. How clever of the ancients to grind the wheat and make it into bread and pasta. How less than clever of modern man to use it to clog plumbing.

We ate simply. We may have lost weight, but we got by. Looking back now I can't remember any complaints. Complaints often come with affluence when the choices being made are not the ones that you would make. During a time of lack, the fact that there is food is enough. Complaints can also come with comparison, when some group of people has more or better than you do. When everybody is in it together, we weather it together – no complaints.

May we and all beings enjoy the blessing of this food we share.

Recipes

One of the foods I turn to to see me through is beets. When wandering the market bereft, I hear the beets, deep, dark, purple red, calling out to me, "We are here for you." Beets, I feel, have extracted the blood of the earth – made dirt red, warm, and flowing. Eating beets is like receiving a transfusion.

Others have noticed this, too, how working with beets is a blessing, which in a literal sense means to be sanctified with blood. I saw one friend shortly after she emerged from the kitchen. Her hands, even after washing, were glowing with tell-tale rose, and she was beaming the praises of beet. Peeling them, she said, is "unbelievably sensual."

Beets are not glamorous or exciting. Their curves, though ample and full, are not smooth, sleek, or beguiling. The greens that top them are sturdy and strong, rather than perky or exuberant. We don't turn to beets for fun and good times, but like a friend or a companion, they will be there, trustworthy and stable. "Thank you," we say. "I know you are there for me."

About Cooking Beets
Beets and Beet Greens with Red Chard
Beet Soup with Red Potatoes
Baked Beets with Dried Cranberries and Sun-Dried Tomatoes

About Cooking Beets

I prefer to cook beets whole with their skins on, and then to peel them. They can be either steamed or baked. Cut the stems off about an inch from the beets, then the beet juices will stay inside the beet, and you will have a little handle to start the skin removal later. If they are fresh and in good condition, you can save them to cook later (one recipe here calls for them), but they can also be added to other vegetable dishes or soups, like spinach or chard.

Put the beets in a saucepot with ½ inch or so of water, which doesn't need to cover the beets. Cover the pot and heat to boiling, then reduce the heat and continue cooking (steaming) until the beets are fork-tender. This will take about 30 minutes, depending on the size of the beets. If the biggest ones are cut in half, they will cook more quickly. Check the pot on occasion to make sure it still has water in it.

Remove the beets from the pot or pan and let them cool for several minutes until you can handle them easily. Remove the stems, and then slip off the skins by hand. Often this is done by squeezing the beet so that the smooth, slippery-surfaced inside pops out.

Cut in the desired shape: rounds, half-rounds, wedges, quarters, or eighths. Notice I do not say "cubes," because I am not a fan of making beets look like they came out of a machine.

Baking the beets will give them a somewhat sweeter, roasted flavor. Put the unpeeled beets (with the inch of stalk) in a baking pan with ½ inch of water, and cover with foil or a lid. Bake at 375 degrees until fork-tender, about 1 hour. As with the steamed beets, wait for them to cool before peeling.

Beets and Beet Greens with Red Chard

Beets with beet greens is pretty basic – a dish with integrity and character. I appreciate its fundamental nature, and I find the combination with red chard quite pleasing. However, if red chard is not available, you are certainly welcome to do without it, and increase the quantities of the other ingredients.

Serves 3 to 4 people

Greens of 2 beets
½ bunch red chard (or the whole bunch if no beet greens are available)
2 medium to large beets (or more smaller ones), cooked
1 tablespoon olive oil or butter
1 red onion, sliced
Salt
Red chili powder (page 19), cayenne, or black pepper
Juice of ½ lemon or rice vinegar

Rinse off the beet greens and the red chard. Beet greens tend to be especially sandy, so immersing them in a sink or bowl of water will be efficacious. Slice the beet greens crosswise into 1-inch sections, including the stems, which become surprisingly tender when cooked. Cut the stems of the chard crosswise into ¼-inch pieces. Cut the largest chard leaves in half lengthwise, then all the chard crosswise into 1-inch sections. Peel the cooked beets and cut them into wedges.

Heat a large skillet with the olive oil, and then sauté the red onion for several minutes. Add the chard stalks and continue cooking a few minutes longer. Then add the beets, beet greens, and chard leaves. Sprinkle with 2 or 3 pinches of salt and a half teaspoon or more of chili powder – be much more modest if you are using cayenne pepper. Cover with a lid, and cook over moderate heat until the greens are tender, 3 to 5 minutes at most.

Season with the lemon juice, adding a little at a time, and see how you like it.

Beet Soup with Red Potatoes

When people plan to make beet soup, they most often think of borscht. While many borschts are quite excellent, I make another kind of beet soup, one that emphasizes the beetiness of beet. Blended to smoothness, this soup is straightforward and soothing.

Serves 4 to 6 people

2 tablespoons olive oil
1 yellow onion, sliced
2 stalks celery, sliced
1 carrot, diced
2 medium red potatoes, diced
Salt
Black pepper
4 to 6 cups water
2 medium beets, cooked and sliced

Optional:
Fresh herbs, minced, for garnish
Thinly sliced lemon

Heat the olive oil in a large skillet and sauté the onion for 2 to 3 minutes until it begins to turn translucent. Add the celery, carrot, and potatoes, and continue sautéing, while stirring, for several minutes longer. Then season with salt and pepper, add 1 cup of water, cover, and cook over low heat until very soft, 10 to 15 minutes.

Add the sliced beets and 2 more cups of water, then puree (in a food processor or with an immersion blender) until smooth. Return to the stove and add more water until you have the consistency you desire.

Check the seasoning and serve. It will also keep well and can be easily reheated later.

Some freshly minced herbs – parsley, thyme, mint, marjoram – could be used as a garnish, or a thin slice of lemon, or both.

Baked Beets with Dried Cranberries and Sun-Dried Tomatoes

This recipe turned out to be "alchemical," which is to say that the flavors unite and blend in a surprising way. Dried cranberries (which are somewhat sweetened or they would be really tart) are becoming widely available.

Serves 4 to 6 people

12 to 15 sun-dried tomatoes
½ cup dried cranberries
5 to 6 beets (probably 2 bunches)
Peel of half an orange, grated
Salt

If you are using sun-dried tomatoes that are not packed in oil, place them in a saucepan, cover with water, and cook several minutes. Add the cranberries for half a minute at the end. Drain and reserve the liquid. Slice the sun-dried tomatoes into narrow strips.

Trim the stems off the beets and place them in a baking dish with the water from the sun-dried tomatoes (or add enough water to fill up the pan ¼ inch). Cover the pan and bake in a 375- to 400-degree oven for 1 hour or more until they are fork-tender.

Remove beets from pan and allow to cool enough so that you can remove the skins, roots, and stems. Then cut up the beets and put them back in the baking dish with the tomatoes, cranberries, orange peel, and a touch of salt. Reheat in the oven before serving.

An Offering

What is it that brings people back, again and again, to the task of feeding, whether the work be a drudgery or a joy? Back to dreaming up what to cook and how to cook it? Back to agonizing over or delighting in what to serve? And back to wondering whether the results are praiseworthy and whether those eating are sufficiently appreciative of all the sacrifice?

Years ago at Tassajara, we had a festive picnic in the early afternoon. I remember walking over the hill to the Horse Pasture. A great deal of effort had gone into the preparations, and people ate eagerly and with gusto. Walking in the fresh air stimulated our appetite for good food, laughter, gaiety, companionship.

Sometimes one attains, for a few moments at least, a heavenly state: sunshine, grasses, wild flowers in bloom, the fleeting and buoyant fragrances of spring, and food. Without having to do a single thing food appears, miraculously, as though borne on the wind. Even before the wish is made, everything is there.

As often happens when things come unasked for, both pleasant and painful, people did not think to say "thank you" that day. Joy, ease, well-being arose, and everyone was replete and sated, and not especially interested in lining up to thank the cooks, the gods, or Divine Providence. Oh well. The euphoria of the cooks on a bright spring afternoon was edged with bitterness: Maybe next time we won't work so hard, if you cannot express more appreciation.

Later in the afternoon I was back in the kitchen at the appointed time. No one else appeared. The other cooks had joined in the general day off, ignoring the well-known dictum that cooking never stops, the kitchen never closes. Even though we'd had a large picnic lunch, people would be expecting a little something when the dinner bell rang.

Unthanked, and now abandoned by my crew, I had little sympathy for what anyone might be expecting. I was, after all, expecting gratitude and plaudits. I was expecting assistance. If you're not going to notice that I'm cooking, let's see if you notice that I'm not. Why don't I disappear? That will show them.

But will it? What will it show them? I fear they will not understand that it's all their fault for not thanking me enough. Before I can get out of the kitchen, further reflection sets in. Do I just give to get? That's not really giving then, that's called buying; and trying to work a favorable deal, that's called bargaining. I want to be more generous than that. I want to really give, no strings attached.

Could I do this, no strings attached? Just cook, and let it go at that? I began sorting through the leftovers. Dinner at the usual time.

The cook's job is to embody generosity, just as it is the work of the people eating to be grateful, even if wordlessly. Still cooks survive better when they focus on their own endeavor and don't try to tell others how they are supposed to react. Cook, and let it go at that. To wit, a story.

In our meditation tradition we have a custom which for many years I found peculiar — offering food to Buddha. Before breakfast and lunch the cooks make up a small tray of food. In its way it is rather cute, suggestive of dollhouse cuisine. Food is put delicately into each little dish, and then a miniature spoon and chopsticks are set ready to use.

When I was cooking, I found this rather annoying. Wasn't I busy enough serving food to the community without having to serve food to someone who's not even going to eat it? You have to be kidding. Whatever is the point?

Later, when I collected the uneaten food, the Buddha didn't say anything about liking this or not liking that, no "Loved the seasoning on the carrots." No, the Buddha just goes on sitting there completely unconcerned. Good food the Buddha doesn't praise. Bad food the Buddha does not complain about. Not a word escapes his lips just as no food escapes his bowls. "How inane," I thought.

When Jakusho Bill Kwong came to visit the first summer I was cooking at Tassajara, I watched him make up the offering tray. Jakusho teaches now at the Sonoma Mountain Zen Center, and he had been the cook at Zen Center when I first arrived. I could not believe how polite and respectful he was while putting food into those tiny bowls: careful, sincere, unhurried, as though serving the most honored of guests, "Please, try this. I'm sure you'll like it." How sweetly he served the food which was to be uneaten and unremarked upon.

Perhaps twenty-five years later it occurred to me that serving food to

Buddha in this fashion was utterly profound: This is the way to cook. Cook the food and serve it. Bow and depart. You've done your part. Offer what you have to offer, let go, and walk away.

That's the end of it. How the guests receive it is up to them.

Recipes

Now I offer some recipes for summer picnic salads and leave it up to you what to do: Pick them up and use them or set them aside. And for your consideration I have some additional preferences: I like my salads at room temperature with the vegetables lightly cooked (rather than raw). I feel this improves their flavor and makes them somewhat easier to chew. I also prefer not using mayonnaise, since I delight in good olive oil and flavored vinegars, as well as the unadulterated colors of the ingredients.

Potato Salad with Corn and Red Pepper
White Bean Salad with Olives and Zucchini
Macaroni Salad with Tomatoes, Bell Pepper, and Red Onion

Potato Salad with Corn and Red Pepper

Serves 4 to 6 people

2 ears of corn (about 1½ cup kernels)
1½ pounds red potatoes, cubed
1 red bell pepper, quartered lengthwise, then cut crosswise into thin strips
¼ cup olive oil
4 to 6 green onions, thinly sliced (white and green parts)
2 to 3 cloves garlic, minced

2 tablespoons tarragon vinegar

Salt

Black pepper

½ bunch flat-leaf parsley, whole leaves removed from stems *or* 1 bunch
    watercress, large stems removed

I suggest cutting the kernels off the corn while it is laying flat on the counter. Often people want to do this by setting the ear of corn on end, and the sweet little kernels delight in bouncing all over the place.

Cook the potato cubes in lightly salted water until tender, about 7 to 8 minutes. When done to your liking, add the corn for about half a minute, then drain (reserving the liquid if you have another use for it).

Meanwhile, sauté the red pepper strips in the heated olive oil for a minute or so, add the green onion and garlic, and continue cooking another minute or two.

Combine the cooked peppers (oil, onion, garlic) with the drained potatoes and corn. Season with the vinegar and salt and black pepper to taste. Toss with the greens and serve.

White Bean Salad with Olives and Zucchini

Serves 4 to 6 people

½ cup white beans (navy beans), (about 1½ cups cooked)

4 cups water

1 large or 2 smaller zucchini

2 shallots

1 tablespoon olive oil

½ large carrot, grated

2 to 3 tablespoons lemon juice

3 ounces Niçoise (or Kalamata) olives, pitted (about ⅓ to ½ cup olive pieces)

1 jalapeño chili, minced

½ teaspoon sugar

Salt

Fresh basil sprigs for garnish (optional)

If you think of it, soak the white beans overnight. Cook with the 4 cups of water until tender, 20 to 30 minutes if beans are presoaked, about an hour if not (or pressure-cook).

Cut the zucchini with the "Chinese rolling cut": Make a diagonal slice, then rotate the zucchini and make a second diagonal slice that partially overlaps the first one. Continue in the same manner. Mince the shallots. Sauté the shallots and zucchini in the olive oil for a minute or two, then add the carrot. After another minute, add the lemon juice, then remove from the heat.

When the beans are tender – completely! – drain them, and combine them with the sautéed vegetables. Add the olive pieces, minced chili, and sugar, then salt to taste.

If you want basil, remove the leaves from the stems, roll a pile of them into a log, and slice crosswise into narrow pieces. Wait for the salad to cool, then mix in the basil. I thought this was also excellent without the basil, but basil is such a big part of summer. . . .

Macaroni Salad
with Tomatoes, Bell Pepper, and Red Onion

Serves 4 to 6 people

½ pound macaroni or shell pasta

1 red onion (about 6 ounces), thinly sliced

1 green bell pepper, quartered lengthwise, then sliced crosswise

2 tablespoons olive oil

1 pint basket cherry tomatoes, halved

½ tablespoon red wine or balsamic vinegar

½ teaspoon red chili pepper *or* Tabasco, to taste

Salt

2 tablespoons capers (optional)

Cook the macaroni in boiling salted water until tender. Add the red onion and green bell pepper slices and continue cooking for another half minute. Drain (reserving the liquid for another use if possible).

Combine with the olive oil, tomatoes, vinegar, red pepper, and salt to taste. If you're in the mood, add some capers.

Attaining Intimacy

Suzuki Roshi Breaks a Tooth

One day when I was the cook at Tassajara, I heard that Suzuki Roshi had broken a tooth while eating black beans. A stone which hadn't been removed from the black beans did the damage. He was going to have to return to San Francisco to have his tooth repaired. Several people came to tell me what an awful person I was for letting this happen.

I should be more careful, they said. I should make sure the crew was more conscientious. We would be missing some of his lectures, they said. He was getting older, and for us to miss any talks was a terrible loss, which could not be made up. I had been horribly negligent. There was simply no excuse.

I have not always had much perspective in situations like this. Many of us don't. I tend to take on the burden of what people say about me. "Yes, you're right," I thought, "I am the worst person ever, causing our beloved teacher to break a tooth. Perhaps I should resign from my position and never cook again. Perhaps I should do penance for the rest of my life."

Even though it was just a broken tooth, I felt remorseful, quite out of scale with the circumstances. The Roshi left to have his tooth worked on, and I had to continue my job, feeling that I was an object of scorn: "There he goes, the person who has caused us to miss out on the precious teaching. It's all his fault."

After a couple days of this I got to thinking that I couldn't be the *sole* cause for what had happened. Suzuki Roshi's teeth had been around from before I was even born, and what, after all, had gone into them? I started thinking about his stories about his life during the Second World War in Japan.

Unlike many other Zen priests who became part-time teachers or took up other occupations to make ends meet during that time of hardship, Suzuki Roshi had decided to be simply a priest. He wanted, he said, to fulfill Dogen-Zenji's injunction that priests not work, but allow the universe to provide.

To this day I am in awe of this decision – leaving his life in the hands of others, especially under such difficult circumstances. He would not even ask others for donations, but wait for someone to notice his need. Perhaps, finally, a parishioner would be visiting his temple and realize with some surprise that he had no food. "I could get you a little rice," the visitor would offer. "I could spare some tea," someone would say. If no one noticed, he simply waited.

"A lot has gone into making his teeth the way they are," I thought. "He must have gone months with rather poor nutrition. His teeth cannot be in the best of shape." After a while I felt better. I didn't feel as much to blame, and I admired my teacher for his willingness to place himself in our hands, eating our food, joining us in our life.

He simply ate what we put in front of him, and never said that we should be offering him foods to which he was more accustomed: white rice instead of brown, or miso soup instead of lentil. He accepted the food we served and other of our American strangenesses, and bowed in respect and gratitude. He let us flounder around while trying to find our way.

When Suzuki Roshi returned from San Francisco with his tooth repaired, students at Tassajara wanted to be sure nothing like this would happen ever again. And it wasn't just a matter of making sure there were no rocks in the beans. They wanted to be sure that the food was soft enough for him, so we went to some extremes endeavoring to provide soft food.

First of all, we tried serving the Roshi a special tray of food before we brought in the regular pots to serve everyone. He refused to accept it. He made it clear that he would eat what everybody else was eating. Then we started cooking some of the food longer or cutting it finer, and putting it in a particular place in the serving bowl where the server would know to serve it to him. We had an uneasy truce with our teacher as kitchen workers outdid one another to provide easily chewable food.

Finally I felt we might be getting a bit extreme, so I decided to speak with Suzuki Roshi himself about what would be appropriate food, considering the condition of his teeth. I explained that we were concerned about his health and well-being and wanted to make sure that the food was soft enough for him. How had that been working for him, I wondered.

"You have no idea," he replied, "how humiliating it is to be served mashed banana."

"Perhaps we have overdone it," I conceded, "but is there anything in particular we should be careful about?"

"No," he said, "I want to eat what everyone else is eating. I will eat whatever you serve."

"But aren't some things more difficult for you than others," I persisted. "What about raw apple slices?"

"No, something crunchy like apple is not a problem. What is more difficult is something chewy like undercooked eggplant, or something stringy like celery."

After that we stopped trying to provide super-soft food for our teacher and started trying to provide everyone with well-cooked eggplant and thinly sliced celery.

It is said that a Bodhisattva's path does not vary. Certainly Suzuki Roshi's path did not waver – from rotten pickles to mashed banana, he accepted what was offered. Perhaps it cost him days and weeks of his life, but then again if he didn't practice the Bodhisattva's way, his spirit might have died much sooner. He joined his life to ours, and left his fate in our hands.

Recipes

We've been eating black beans for years at Zen Center, often in the form of Black Bean Chili, which is frequently served with corn bread and a green salad. While I don't necessarily always make the whole menu, I certainly enjoy each of the dishes.

Black Bean Chili
Corn Bread
Garden Salad
Baked Goat Cheese

Black Bean Chili

This is a simplified version of the recipe in The Greens Cookbook *— more like what I do at home.*

Serves 3 to 4 people

1 cup black beans, soaked overnight if possible

1 bay leaf

2 teaspoons cumin seeds

2 teaspoons dried oregano

¼ teaspoon cayenne

2 to 3 tablespoons ground chili (see page 19)

1 tablespoon olive oil

2 small or 1 large yellow onion, diced or sliced

2 cloves garlic, chopped

¼ teaspoon salt

1-pound can unseasoned tomatoes, chopped

1 teaspoon balsamic vinegar

2 tablespoons minced cilantro

Garnishes:

4 to 6 ounces BruderBasel or other smoked cheese, grated

Sprigs of cilantro

Sort through the black beans so that you can remove any small stones. Black beans are notorious for this. Soak the beans in water overnight or during the day. I usually use the soaking water to cook the beans, but you can drain that liquid if you like and replace it. Some people think that the beans will produce less "gas," if the water is replaced.

Cover the beans with 2 inches of water and cook with the bay leaf about 30 to 40 minutes if you have soaked them and about 75 to 90 minutes if you have not, until they are completely tender.

Grind the cumin seeds, combine with the oregano, cayenne, and ground chili, and roast in a dry pan for several minutes over moderately low heat until fragrant, stirring as necessary.

Heat the oil in a saucepan, and sauté the onion for a few minutes until it becomes translucent. Add the garlic, seasonings, and salt and continue cooking another 2 to 3 minutes over low heat. If the mixture begins to stick, add some water from the bean pot. Add the tomatoes and cook until heated through, stirring the seasonings off the bottom of the pot. Add to the beans once they are tender. Adjust seasoning, adding the vinegar and anything else to spice it up to your liking.

The *Greens* version makes a fine presentation which calls for Muenster cheese in the bottom of each bowl, then the chili, and finally sour cream, green chilies, and cilantro on top. At home I am more likely to garnish with the grated smoked cheese and a few cilantro sprigs.

Corn Bread

½ cup coarse cornmeal (polenta)

1½ cups white flour

4 teaspoons baking powder

½ teaspoon salt

½ cup sugar

1 cup milk

3 eggs, beaten

⅓ cup melted butter

Preheat oven to 375 degrees.

Mix together the cornmeal, flour, baking powder, salt, and sugar. Combine the milk, eggs, and butter. Fold together the wets and dries, mixing just enough to combine, leaving perhaps even a few dry places. Overmixing will make the corn bread tough. Scrape batter into a greased 9 x 13-inch baking pan. Bake in a 375-degree oven for 25 to 35 minutes until the middle rises and top is cracking here and there.

Let cool briefly and then cut into squares to serve.

Garden Salad

This is called "Garden Salad" because I go out to my garden and pick some greens for salad.

Often I get my lettuce starts from Green Gulch, Zen Center's farm near Muir Beach. Here they have many wonderful varieties of lettuce from seeds that have been prized and shepherded: Marvel of Four Seasons, Black-Seeded Simpson, Kagraner Somer, Lolla Rosa, Cool Mint Romaine, Red Leaf Romaine, Red and Green Oak Leaf lettuces.

Since I have been growing lettuces in my yard, I find it difficult to buy lettuce. The store-bought heads do not seem to have the hardiness and vigor of the leaves that I harvest by hand. After I have picked an assortment of lettuce leaves, I poke around for what else to add: spearmint, thyme, arugula, parsley, lime thyme, perhaps some baby chard or mustard greens, some calendula blossoms or nasturtium flowers (or leaves).

Nowadays markets frequently have available a mix of fresh greens (often referred to as mesclun*). Some are better than others, so you may need to experiment to find out which particular mix you enjoy most.*

Serves 4 people

¾ pound lettuce of your choice

1 to 2 ounces or more miscellaneous garden herbs or flowers (See list above or
   look in your own garden.)

1 shallot, diced *or* 1 small bunch chives, chopped

1 tablespoon olive oil

2 teaspoons balsamic vinegar

Salt

Black pepper

Optional:

1 or 2 tablespoons sesame or sunflower seeds, roasted

2 or 3 tablespoons Asiago or Parmesan cheese, freshly grated

Pick your lettuces and herbs. Wash by immersing in water, then spin dry in a lettuce spinner. Tear the lettuce into fork-sized pieces. For this salad I tend not to mince or chop the herbs, but leave them in whole-leaf form removed from their stems.

Toss the greens with the shallot and olive oil. If everything is not quite moistened enough with the oil, add a touch more. Then toss with the vinegar and a sprinkling of salt and pepper.

As an alternative to the balsamic vinegar I might toss with a tablespoon of rice vinegar and a few pinches of sugar.

For variation, garnish with the roasted seeds or grated cheese.

Baked Goat Cheese

I like to keep a package of fresh goat cheese available. I didn't always like goat cheese (chèvre), but now there are many fine fresh California goat cheese producers, along with the imported French ones. This is a soft cheese which can be crumbled or spread, or in this case, baked.

Serves 2 to 4 people

1 (5-ounce) package of goat cheese (chèvre)

1 teaspoon olive oil (optional)

2 teaspoon fresh herbs, minced (parsley, thyme, oregano) *or* dried *herbes de Provence*

Garden Salad (see recipe page 188) *or* suitable vegetable dish

Preheat oven to 350 degrees.

The baked goat cheese is usually served on a bed of salad greens, but could also be served over other vegetable dishes, possibly the Broccoli with Olives and Lemon (see recipe page 156) or Beets and Beet Greens with Red Chard (see recipe page 172). It seems especially good with greens, fresh or cooked.

Divide the goat cheese into 2 to 4 portions and shape them into small patties. Coat with olive oil if using it. Then roll in the herbs.

Place on a baking sheet and bake in a 350-degree oven for 5 minutes. Serve on the dressed salad greens or cooked vegetables.

VARIATIONS

I also like rolling the goat cheese patties in roasted sesame seeds or roasted, chopped almonds or walnuts. So good!

Making the Perfect Biscuit

When I first started cooking, I had a problem: I couldn't get my biscuits to come out the way they were supposed to. I'd follow the recipe and try variations, but nothing worked. I had in mind the "perfect" biscuit, and these just didn't measure up.

Growing up I had "made" two kinds of biscuits: One was from Bisquik and the other from Pillsbury. For the Bisquik biscuits you added milk to the mix and then blobbed the dough in spoonfuls onto the pan – you didn't even need to roll them out. The biscuits from Pillsbury came in a kind of cardboard can. You rapped the can on a corner of the counter, and it popped open. Then you twisted the can open more, put the premade biscuits on a pan, and baked them. I really liked those Pillsbury biscuits. Isn't that what biscuits should taste like? Mine just weren't coming out right.

It's wonderful and amazing the ideas we get about what biscuits should taste like, or what a life should look like. Compared to what? Canned biscuits from Pillsbury? "Leave It to Beaver"? And then we often forget where the idea came from or that we even have the idea. Those (perfectly good) biscuits just aren't "right."

People who ate my biscuits could be extolling their virtues, eating one after another, but for me, they were not "right." Finally one day a shifting-into-place occurred, an awakening: not "right" compared to what? Oh, no! I've been trying to make canned Pillsbury biscuits! Then that exquisite moment of actually tasting my biscuits without comparing them to some (previously hidden) standard: wheaty, flaky, buttery, "sunny, earthy, here." Inconceivably delicious, incomparably alive, present, vibrant. In fact, much more satisfying than any memory.

Those moments when you realize your life as it is is just fine, thank you, can be so stunning and liberating. Only the insidious comparison to a beautifully prepared, beautifully packaged product makes it seem insufficient. The effort to produce a life with no dirty bowls, no messy feelings, no depression, no anger is bound to fail – how endlessly frustrating.

Sometimes when I am cooking, Patti asks if she can help. My response is often not pretty, neat, or presentable. The lid comes right off, and I blow it: "No!" How could an offer of assistance be so traumatic and irritating? Neither of us could understand how my response could be so out of scale, so emotionally reactive. But I suppose it just depends on which biscuit you're trying to bake.

I couldn't get it for the longest time. While I was no longer trying to be the greatest chef ever, I realized that I was still trying to make myself into the Perfect Grown-up Man: competent, capable, and superbly skilled, performing every task without needing any help.

Someone's asking, "Anything I can do?" implies that I need help, that I somehow am not competent, independent, and grown-up enough to handle the cooking myself. Ironically, the desperate attachment to being the Perfect Grown-up meant becoming a moody, emotional infant with strange pricklinesses. "How could you think such a thing?" I would rage. "You've ruined my Perfect Biscuits. Now leave me alone!"

As a zen student one can spend years trying to make it look right, trying to cover the faults, conceal the messes. Everyone knew what the Bisquik Zen student looked like: calm, buoyant, cheerful, energetic, deep, profound. Our motto, as one of my friends says, was "Looking good."

We've all done it: tried to attain perfection; tried to look good as a husband, wife, or parent. "Yes, I have it together." "I'm not greedy or jealous or angry." "You're the one who does those things, and if you didn't do them first, I wouldn't do them either." "You started it."

"Don't peek behind my cover," we say, "and if you do, keep it to yourself." Well, to heck with it, I say, wake up and smell the coffee. How about savoring some good old home-cooking, the biscuits of today?

Recipes

In my boyhood Sunday mornings were the time I would endeavor to make biscuits or coffee cake. Now it's still on Sunday morning that I get around to baking.

New Flaky Biscuits
Ginger Muffins
Breakfast Custard

New Flaky Biscuits

Makes about 1 dozen biscuits

1 cup whole-wheat flour

1 cup unbleached white flour

¼ teaspoon salt

½ teaspoon baking soda

½ teaspoon baking powder

½ cup butter

½ teaspoon vanilla extract

1 egg

½ cup plain yogurt

Sugar or cinnamon sugar (optional)

Preheat oven to 475 to 500 degrees.

Combine the flours, salt, soda, and baking powder, and cut in the butter until it is in small lumps. Make a well in the center and put in the vanilla, egg, and yogurt. Mix these together with a fork, then lightly, minimally mix in the flour. (Overmixing will make the biscuits tough.)

Roll out on a floured board to about ½ inch thick. Fold in thirds. Repeat this twice. Then roll out to ½-inch thickness, and using a biscuit cutter, glass, or cup, cut into biscuits. Dip the cutter in flour, if necessary, so that the biscuits will not stick to it. Place the cut biscuits on an ungreased baking sheet. Gather up the remains and reroll to make more biscuits.

Sprinkle the tops with sugar or sugar mixed with cinnamon for a touch of sweetness.

Bake on the top shelf of a very hot oven, 475 to 500 degrees, for 8 to 10 minutes. The biscuits are done when they are brown on the *bottom*. Lift one up to check. Don't wait for them to get brown on top; then they will be too dried out.

Ginger Muffins

1 cup unbleached white flour

1½ cups whole-wheat flour

¼ teaspoons salt

2 teaspoons baking powder

½ teaspoon baking soda

2 eggs

1½ cups plain yogurt

½ cup honey

Zest of 1 orange

⅓ cup canola oil (or safflower or corn)

2 tablespoons fresh ginger, grated

Nutmeg

Preheat oven to 375 degrees.

Combine flours with salt, baking powder, and soda. In a separate bowl mix together the egg, yogurt, honey, orange zest, oil, and fresh ginger. Pour wets into dries, and mix with as few strokes as possible (20 stirs is usually about right). I use a rubber spatula for this, so I can get the batter off the sides. Over-mixing will make the muffins tough.

Grease a muffin tin (unless you have one of those nice, new nonstick ones). Spoon in the batter to near the top of the muffin cups. Grate some nutmeg over the tops of the muffins.

Bake in a 375-degree oven for about 30 minutes, until the tops have rounded and cracked, and the sides and bottoms have browned. On a recent batch I discovered once again that it is not a good idea to use the bottom shelf of the oven, as the bottoms of the muffins will tend to blacken.

Breakfast Custard

Once in a while I have planned ahead and made a Breakfast Custard the evening before. Then it is ready in the morning to have with the Flaky Biscuits or the Ginger Muffins. The soft, smooth, light texture of the custard can be quite comforting.

Serves 3 to 4 people

3 eggs
¼ cup brown sugar
Pinch of salt
½ teaspoon vanilla extract
¼ teaspoon cinnamon
2 cups warmed milk
Pot of boiling water

Preheat oven to 325 degrees.

Whisk the eggs. Whisk the sugar, salt, vanilla, and cinnamon into the warmed milk. Then whisk the milk into the eggs.

Pour into four 1-cup baking dishes (or 1 large ceramic or glass baking dish). Place in baking pan, then set in preheated 325-degree oven. Pour hot water into baking pan so that it is half way up the sides of custard dishes. Bake 50 to 60 minutes or until the custard is firm.

Note that the custard will take much longer to bake if you do not preheat the milk, if you do not start with boiling water, or if you bake it in 1 large dish.

When baked the night before, the custard is ready and waiting for you when you get up in the morning. I prefer it at room temperature.

Serve the custard accompanied by fresh fruit, especially summer fruits such as berries, cherries, apricots, plums, peaches, and nectarines, with the larger fruits cut up into slices and sprinkled perhaps with some Rose-Scented Sugar.

Feeling Your Way Along

Finding out how to cook or how to work with others is something that comes with doing it, feeling your way along. And the more you master your craft, the more you know that the way is to keep finding out the way, not by just doing what you are already good at, but by going off into the darkness.

My teacher, Suzuki Roshi, once emphasized this point during a week of intensive meditation: "Zen," he said, "is to feel your way along in the dark, not knowing what you will meet, not already knowing what to do.

"Most of us don't like going so slowly, and we would like to think it is possible to figure everything out ahead of time, but if you go too fast or are not careful enough, you will bump into things. So just feel your way along in the dark, slowly and carefully," and he would gesture with his hand out in front of him, feeling this way and that in the empty air.

"When you do things with this spirit, you don't know what the results will be, but because you carefully feel your way along, the results will be okay. You can trust what will happen."

Following the week of meditation at the ceremony where all the students ask the teacher a question, I walked forward with my hand outstretched, investigating the air. "Feeling our way along in the dark," I began, "and now that sesshin is over, can we have a party?"

"If you really do it with that spirit, then it will be OK," he answered. I bowed and said, "Thank you," and started to leave, but then his voice began again. "The most important point," he said and paused, while I came to a complete stop, focusing intently on what he would say, "is to find . . . out. . . ." Again he paused as I awaited the conclusion. "What . . . is . . . the most important point."

He had drawn it out so I was hanging on every word: To find what? To find out what? To find out what is what? And then he hadn't really told me the most important point, only he had. I was momentarily disappointed: "But I thought you were going to tell me the most important point." Then I felt a kind of joy or elation: "Yes, I could be finding out what is the most important point."

So I began investigating "what is the most important point." One day I would feel it was sincerity, but other days it might be gratitude, equanimity, or letting go. Moments of experience seemed to brighten: tasting intently, speaking kindly, slowing down, being present, finding pleasure and ease in the work of cooking.

Adverse situations also became occasions for trying to find the most important point. When things are not working, what's important then? Compassion, patience, good-heartedness, not being so hard on yourself, clarifying your innermost request and acting on it. One thing after another seemed pivotal.

Finally after some years of study, I concluded that the most important point is to be finding out what is the most important point. Doing that, after all, had given me a focus beyond the urgencies of everyday life, a focus which made it more possible to extract the nutritive essence from each experience.

Discovering how to cook or how to plan menus; learning how to adapt to space limitations and time restrictions, how to handle ingredients – all these were feeling my way along, finding out what is the most important point.

Recipes

I never much liked cooked cabbage until my mom discovered a recipe where it was steamed in California sauterne – which is an inexpensive table wine quite unlike the French dessert wine it is named after. My mom was good at feeling her way along until she found something that worked. I still enjoy Drunken Cabbage as a side dish, and I have a second version for you as well.

Also, feeling my way along, I've come up with a cabbage "lasagna."

Drunken Cabbage
Red Cabbage with Sake and Green Onions
Cabbage Lasagna

Drunken Cabbage

This is a bare-bones recipe, so if you want to add garlic, ginger, red pepper, herbs, spices, it's OK, as long as you feel your own way along and do not count on me.

Serves 4 to 6 people

1 tablespoon olive oil
12 to 16 ounces green cabbage, sliced in shreds
1 cup white wine
Salt and pepper

Heat a large skillet, add the olive oil, then the cabbage, and sauté for 2 or 3 minutes, stirring. Then add the wine and season with salt and pepper. Cover and steam a few minutes until the cabbage is tender. Check the seasoning. That's it.

VARIATIONS

Before adding the cabbage, sauté some sliced onion and carrot for a fuller-tasting, more colorful dish. Garnish with parsley or green onion.

My editor, Jisho, suggests chenin blanc for a light, clear flavor, or pink zinfandel for a sweeter finish.

Red Cabbage with Sake and Green Onions

I find the flavors of this variation to be quite enjoyable.

Serves 4 people

6 to 8 green onions
1 tablespoon olive oil
2 teaspoons fresh ginger, grated
12 ounces red cabbage, sliced in shreds
1 cup sake
Salt
Black pepper

Attaining Intimacy

Slice the green onions on fairly thin diagonals, setting aside the dark green parts to add later. Heat a skillet, add the oil, then the whites and pale green parts of the onion. Cook for minute or so and add the ginger. After another minute, add the cabbage. Stir-fry for a couple of minutes, then add the sake and a sprinkling of salt and pepper. Cover, lower the heat, and cook until fairly tender. Add the dark green of the green onion, and let it get limp. Check the seasoning and serve.

Cabbage Lasagna

I do not recall where I originally got the idea for this, but I make it periodically with ingredients I happen to have around. It is easy to prepare and surprisingly delicious. Most recently, I made this with some leftover Ancho Chili Sauce and two cheeses my daughter Lichen had brought back from France: Le Frommage Basque and Chaource. Fantastique! Here's the picture – another chance to feel your way along.

Serves 4 to 6 people

Small head of cabbage (about 1H pounds)

1 teaspoon salt

1 pound fresh tomatoes *or* 2 cups Herbed Tomato Sauce (see recipe page 246)
      or a combination of the two

6 ounces meltable cheese (whatever you have around that you like)

⅓ cup Asiago *or* Parmesan cheese, grated

1 tablespoon fresh thyme, minced, *or* parsley

If using fresh tomatoes:

1 tablespoon balsamic vinegar

2 cloves garlic, minced

Salt

Black pepper

1 teaspoon *herbes de Provence or* ½ teaspoon dried thyme with ½ teaspoon
      dried basil

Preheat the oven to 375 degrees.

The idea of this recipe is to blanch the cabbage, then layer it in a baking pan or casserole with the tomato and cheese.

Start 2 quarts of water boiling. Cut the core out of the cabbage and peel off the individual leaves. If they do not come loose readily, put the head of cabbage in your heating water for a minute or so, then remove to a bowl where it can drain and you can pull off the leaves. Do this as necessary.

The alternative is to cut the cabbage lengthwise into quarters, and then cut out the core. This is easier, but results in the drawback of many more smaller pieces of cabbage.

Once all the individual leaves have been removed, salt the water lightly – using perhaps a teaspoon's worth – and blanch 6 or 8 leaves at a time for about 1 minute. Remove with tongs or a strainer, and drain. Set aside.

If using fresh tomatoes, cut them crosswise into slices and season them with the balsamic vinegar, garlic, salt, pepper, and herbs. (Presumably the tomato sauce will already be seasoned.)

Grate the cheese (or possibly slice it). I tend to have provolone, Gouda, goat, feta, or possibly Jack around. Perhaps you prefer mozzarella, Fontina, Gruyère, Emmenthaler, or Cheddar.

Use a 2-quart casserole if you have one, or a 9 x 13 x 2-inch baking pan, preferably glass-bottomed, as aluminum will react with the cabbage and the tomato. Line the bottom of the pan with a layer of cabbage and distribute a few tomato slices or about ⅓ cup of sauce on top of the cabbage – you will not be covering the cabbage completely. Then distribute about a quarter of the cheese – again you will not be making a complete layer. Repeat the layering: cabbage, tomato, cheese, until all the ingredients are used. Distribute the Asiago or Parmesan cheese on the top.

Bake uncovered at 375 degrees for about 30 minutes until it is bubbling hot. Remove and garnish with the thyme or parsley.

The Benefits of Watermelon

Once a student asked the master, "What about the student who leaves the monastery and does not return?"

"He is a horse's ass," replied the master.

"What about the student who leaves the monastery and later returns?"

"He remembers the benefits."

"What are the benefits?" the student queried.

"Heat in the summer and cold in the winter."

Perhaps this is dry Zen humor, but it is also a wonderful answer, to be reminded that there is a benefit, a simple beauty in the way things are. Not that I always understood or appreciated this when I was actually living at Tassajara. Up to a point, certainly, but heat and cold were often a torment.

At times the usual conceptions of heat and cold were inadequate to describe what we experienced, and so the words lost their meaning. One summer working in the old kitchen I found out how relative heat can be. During one hot spell the temperature was over 115 degrees. This heat met the body head-on. It was no longer just the surrounding environment, but an independent presence, a being that licked at our faces and pushed its body up against ours, took hold of us and squeezed out our vitality.

The kitchen was hotter by ten degrees or more than the baking outdoors. Sweat would pour down our faces and soak through our clothes. We'd walk outside every now and again to cool off. So what was hot? What had been sticky and oppressive a few minutes earlier was now cool and breezy. "Hot" had lost its relevance. Hot compared to what? This morning, last winter? A few degrees less, and it would still be "hot." Twenty degrees less and it would still be "hot."

Everything in the world was the world of heat. Doorknobs felt different – no longer solid, but alive and pulsing. Large stones seemed to breathe. Here and there the air would shimmer and vibrate with reflected heat, making objects look twisted and wavy-edged.

The little voice that liked to whine "If it was just a little cooler . . . then I would be alright," was silenced. Awe took over. Incredulousness. Adventurousness. The response to that dear little voice would come back loud and clear: "If it was just a little cooler, you'd still be hot! so forget it." Here was hot that could not be escaped, hot that transcended hot.

One late summer day Roovan, who was the gardener, and I took advantage of the heat to have a special dinner. Roovan had somehow gotten hold of a large watermelon, and he offered to share it with me.

"Let's have dinner," he said.

"Just watermelon?" I asked.

"Sure," he replied, "what could be better?"

To make it a real dinner though, we agreed to have "courses." We sat outside where we could spit out the seeds in a civilized fashion, and began our meal with some thin slices for hors d'oeuvres, then some wide wedges for "soup."

"Are you ready for the entree?" Roovan wanted to know. "Let's have watermelon steaks."

So we cut big thick rounds for the entree. Sure was delicious: juicy, fragrant, sweet, succulent, that slight crispness which dissolved away into elixir.

Our hands and faces dripped, the rinds piled up. "Now let's have some salad." The pieces got more misshapen. "How about dessert?" "Would you like some coffee?" We cut out lengthy cylinders for after-dinner cigars. "Brandy?"

The world appears in infinite forms and shapes that we could never imagine, often beyond the limits of what we consider comfortable or pleasant. To find benefit in the way things are frees us from trying to make everything conform to our standards. Like watermelon on a hot day – what a relief.

Recipes

How do those melons do it? They can sit in the blazing sun in vast open spaces and grow large, round, ripe, moist, sweet, and juicy. Instead of shriveling up, they sit in the sun and grow fat and full of water. What a remarkable blessing melons provide. They bring me great joy and benefit.

Melon Salad with Lime and Mint
Melon with Tamarind Date Chutney
Melon Platter with Avocado and Fresh Figs

Melon Salad with Lime and Mint

I make this salad with whatever melon I find available: cantaloupe, honeydew, watermelon, casaba, crenshaw or sometimes a colorful combination – such delicious refreshment.

Serves 4 to 6 people

½ pound melon (see above)

Juice of 1 lime

2 tablespoons white sugar or maple syrup

Salt

15 to 18 mint leaves, sliced into narrow strips (about ¼ cup)

Cut the melon open and remove the seeds. If using watermelon, remove the seeds while you work with it. Use a melon "baller" to make melon balls, or cut off the rind, and cut the melon into chunks.

Combine the lime juice and sugar or maple syrup, and toss it with the melon, along with a couple of pinches of salt. Garnish with the mint leaves.

The simplified version, of course, is to slice the melon and serve it with wedges of lime.

Melon with Tamarind Date Chutney

For this dish you'll need to find some tamarind. Use the tarmarind concentrate, if you can get it, since it is all cooked, strained, and ready to use. Lemon can be substituted for the tamarind, if you can't find it.

> 1 honeydew melon
> 1 teaspoon sugar
> 2 pinches of salt
> Tamarind Date Chutney (see recipe below)
> Fresh mint for garnishing

Cut open the melon, remove the seeds, cut off the skins, and cut the flesh into chunks. Sprinkle with the sugar and salt, and set aside.

Put the chutney on bottom of a platter or individual serving plates, then the melon cubes on top. Garnish with mint leaves.

Tamarind Date Chutney

This chutney could also be served with Indian food, and what is left over will keep well in the refrigerator.

> ½ cup tamarind concentrate *or* juice and minced peel of 1 lemon (in addition to
>   lemon below)
> ½ cup water
> 2 tablespoons molasses
> ¼ cup honey
> Juice of 1 lemon
> Pinch or two of ground cumin
> ½ pound pitted dates
> Fresh mint, for garnish

Combine the tamarind, water, molasses, honey, lemon, and cumin and taste. Is it OK so far? Place in a pot and cook with the dates over low heat until the dates can be mashed and stirred in. Adjust the seasoning if needed, especially the ratio of sweet and sour.

Melon Platter with Avocado and Fresh Figs

Three of my favorite foods: one beautiful presentation.

Serves 4 people generously

½ cantaloupe or honeydew melon

2 avocados

8 fresh, ripe figs *or* substitute a second type of melon *or* radish slices

3 tablespoons olive oil

2 tablespoons rice vinegar

1 teaspoon sugar

⅛ teaspoon salt

1 shallot, finely diced *or* ½ bunch chives, finely sliced

2 tablespoons fresh mint, cut into thin strips

Since I do not have a melon "baller," I cut the peel off the melon and then slice it into C-shaped pieces, but if one is available, you could also make melon balls. Prepare decorative avocado slices (see page 64). Cut the stems off the figs, and then slice them in half lengthwise.

Arrange the avocado and fruits on a large platter. A general rule of thumb is to begin with the ingredient that is the largest, either the melon or the avocado, and then fill in with the smaller pieces.

When I don't have figs or a second type of melon to complete the platter, I have found sliced radishes to be an excellent addition, strewn over the other ingredients.

Whisk together the olive oil, vinegar, sugar, and salt. If using shallots whisk them in, as well. Spoon the dressing over the avocado, melon, and figs. If you are trying to use less oil, OK, use more vinegar and sugar.

If using chives instead of shallots, distribute them along with the mint over the top of the arrangement.

I'll Always Be with You

After I had worked in the kitchen at Tassajara for three summers and two winters, I was put out to pasture, a pasture that turned out to have rocks in it. I felt lonely and forlorn, but also unburdened of my identity as a cook. The autumn leaves were dropping. I had been used up and discarded.

First I was asked to work on removing the rocks from the lower garden. We did this by tossing shovelfuls of dirt against a wire screen propped at an angle. The dirt and smaller rocks went through, while the larger rocks remained on the screen. The stones had to be shoveled into wheelbarrows and hauled away.

After the stress of preparing three meals a day, this was really simple, uneventful work. Loosening a large rock from the soil was the big event of the day. Not much progress or accomplishment to speak of: just dig and toss, shovel and haul. No deadlines and no pressure to impress others with my talent or skill. Some days I worked with one or two other people; on other days I had the field to myself. Days turned to weeks, and weeks became months, as the area of sifted soil expanded. Twenty-five years later I still find it gratifying when I walk through the garden.

Once the dirt had been sifted we started building a stone wall on the far side of the garden against the hillside. None of us had done rock work before, but we puttered along, learning as we went. After a time Suzuki Roshi began to notice us, and I told him how much I liked working with rocks.

Later he invited me to do rock work with him around his cabin. Watching his steady pace, unrushed yet intent and absorbed, made it clear to me that even though I had slowed down since leaving the kitchen I was still too fast for rocks, particularly as they got bigger.

The Roshi wasn't in a hurry to get things done. He just kept working at making the rocks fit. Where I would become discouraged or depressed that the rocks didn't fit, he just kept going, tried something else, started over. A large rock that didn't fit after much effort, which would have been a "defeat" for me, was just another event for him. Yet he wasn't a perfectionist either. Where I might have searched for hours or days for a more perfectly fitting rock, he simply wedged in smaller rocks to secure the bigger one or chiseled off rough edges.

He had his eye out for rocks, and knew where each one lived. Once he had us get a large rock forty feet up the hillside above the lower barn. We couldn't understand what he wanted that rock for until we finally got it in place, and it turned into a solid cornerstone. I began to live rocks the way I had lived flavors and menus: Shapes settled into place, pieces fit together.

I also noticed that Suzuki Roshi moved rocks the easy way, utilizing iron bars and stone fulcrums to move larger rocks this way or that with minimal physical exertion and maximum delicacy and exactness. He wasn't impressed by senseless exertion. "Just a moment," he would say, beginning to fiddle with his little curved iron bar.

He meticulously completed each step, especially the finishing work: wedging small stones in behind the dry wall, pounding dirt under paving stones. This is the part of rock work which is most like washing the pots or cleaning the dishes: time-consuming and not particularly creative.

I felt unworthy of his attention, but I appreciated his generous and warm-hearted interest in me. Even though I wasn't always very happy with myself, that seemed OK with him.

Then one day he moved me more firmly into place inside. I hadn't known what he saw in me any more than I had known what he saw in that cornerstone-to-be, so far up the hillside. Yet he watched and waited, and when the time came, got out his lever, inserted it just so, and shifted me.

I was in his cabin, and we had just completed a conversation about some troubling issue I can't recall: Anger, frustration, discouragement, sorrow, grief, abandonment – there were so many. I bowed and got up to leave, and Roshi surprised me by getting up too. Then he walked over to where I stood, put his arms gently around me, and said very simply, "I'll always be with you." And he hugged me, which he had never done before. Tremendous joy and energy soared through me, and I knew it was true: He would always be with me. What had been the solidity of my body was a trembling mass of warm vitality. This incredible being, this pure heart, this kind mind, would never leave me. Ever.

Sometime later I realized he'd always been there.

So with each of us, our sincere heart can be awakened, touched, moved, by a smile or a gesture, by being seen and known, respected and appreciated. And we can do that too: awaken others.

Recipes

Still with me from when I was growing up are Mystery Bites, a cookie bar my mom would make during the holidays. Also keeping me company these days are Chocolate Walnut Cookies and Apple Tart Cake. I don't just eat them; I savor them.

Mystery Bites
Chocolate Walnut Cookies
Apple Tart Cake

Mystery Bites

A holiday favorite in my family; even now I am sometimes inspired to bake these. It's traditional! My mom still has the recipe on a 3-by-5 card "brown with age." I substitute butter for the original vegetable shortening.

   Makes 2 to 3 dozen

½ cup butter

1 cup plus 2 tablespoons unbleached white flour

1½ cups brown sugar

2 eggs

½ teaspoon baking powder

¼ teaspoon salt

1 teaspoon vanilla extract

½ cup shredded coconut

1 cup chopped walnuts or almonds

Preheat oven to 350 degrees.

Cut the butter into 1 cup of the flour and ½ cup of the sugar with 2 knives or a pastry cutter until mealy. Press into the bottom of an 8 x 8-inch baking pan. Bake for 15 to 20 minutes until aromatic and the dough has begun to pull away from the sides of the pan – it does not need to brown.

Beat the eggs, and then mix in the remaining cup of sugar, 2 tablespoons of flour, the baking powder, salt, and vanilla extract. When this is well mixed, fold in the coconut and nuts. Spread over the bottom layer, and return to the oven. Bake an additional 25 to 30 minutes until the top is firm and dry.

Allow to cool, then cut into squares.

Chocolate Walnut Cookies

The original Tassajara Bread Book *did not have a single cookie recipe. Sure, there were bars and squares, but no cookies: I never had time to make them. To roll out individual cookies for fifty or eighty people can take awhile, so I made sheet pans of desserts which I could cut into servings.*

Since then I have rediscovered cookies, and this Chocolate Walnut Cookie is an ethereal one, which simply melts in your mouth. I could not believe how good these were the first time I tried them.

Makes 3 dozen or more cookies

¾ cup unsalted (sweet) butter, softened

¾ cup white sugar

1 egg

4 ounces unsweetened baking chocolate, melted and cooled

2 teaspoons vanilla extract

1 cup finely ground walnuts

1 teaspoon baking powder

2¼ cups unbleached white flour

Preheat oven to 375 degrees.

I usually just dump everything into the bowl and mush together with my hands, but if you want . . .

Cream the butter and blend in the sugar. Mix in the egg, beating well, then the chocolate, vanilla, and nuts. Combine the baking powder with the flour, then mix into the other ingredients.

The dough should be dry enough to shape into balls with your hands without too much sticking to your fingers and wet enough that it does not crumble apart. If necessary, add more flour to make it drier. Add a spot of water or a bit more butter to moisten.

Roll the dough into balls the size of walnuts and place about 2 inches apart on an ungreased cookie sheet. Press them flat with a cookie stamp, a glass, or teacup. Flour the bottom of the pressing implement if it is sticking to the cookies.

Bake at 375 degrees for 8 minutes. They are done when the top of the cookies are cracked, and the bottoms slightly browned.

Apple Tart Cake

This "tart cake" developed from the Turkish Coffee Cake Cookie Bars in my Tassajara Bread Book. I have been working on it for some time now, but perhaps you can still improve on it. The idea is to have a layer of tart dough on the bottom of the pan, then fruit with custard, then cake. But it's easier than that. Check it out.

Makes 1 (8-inch) cake

½ cup whole-wheat flour

1 cup unbleached white flour

⅜ cup white sugar

⅝ cup unsalted (sweet) butter

2 apples, quartered, cored, and sliced

2 tablespoons white sugar

1 egg

¼ cup milk

½ teaspoon cinnamon

1 egg

½ cup sour cream

1 teaspoon vanilla extract

¼ teaspoon baking soda

Preheat oven to 375 degrees.

Combine the flours and sugar, then cut in the butter with 2 knives or a pastry cutter, until it is the consistency of cornmeal. (You will have about 3 cups altogether, loosely measured.)

Press half of the mixture into an 8-inch baking pan. (The bottom will come loose in any case, but it will help later if you have greased the sides.) Arrange the apple slices on top of the crust. Sprinkle with a couple of tablespoons of sugar to sweeten the apples a little.

Now take out a half cup of the remaining crumble mixture (about a third of what is left). Beat 1 egg with the quarter cup milk and the cinnamon, then stir in the half cup of crumble. Distribute over the apple slices.

Beat the other egg with the sour cream, vanilla, and soda. Stir in the remaining crumble mixture, and pour/spread it on top. Bake for about 40 minutes.

The top should be golden brown, and a fork or toothpick inserted in the center should come out clean.

Let cool in the pan. If your cake pan has one of those little lever gizmoes that spins around the bottom of the pan, you can use that to loosen everything. Otherwise use a knife to loosen the sides. Place a plate upside down over the top of the pan, turn everything over, and encourage the cake to place itself on the plate. Then turn right side up onto another plate.

Serve with flavored whipped cream, ice cream, or stewed fruits. Or the way I usually do – plain.

Playing with Fire

I learned something on a rafting trip on the Salmon River in Idaho while watching a companion start a fire with wet wood. The day had started out bright and clear, but by late morning the sky was dark gray and a light continuous rain was falling. Still a long way from the evening's campsite, we had to put in a couple of hours of steady paddling against a cold wind to arrive at our destination. Since the river was wide in that stretch, the current didn't help much. Summer fun sort of thing.

When we stopped briefly at a riverside store and bought candy bars, I realized I had never been so cold and wet in my life. My teeth were chattering and I couldn't stop my body from shaking with uncontrollable spasms up and down my legs.

After another hour of focused effort we reached our campsite. To set up camp we first tied a large tarp to several trees to get a roof over our heads. While some people began assembling tents, often under the cover of the tarp, one of the fellows began assembling the fire with a few scraps of paper, some dry leaves, and minuscule twiglets. Then very thin twigs broken into perhaps three-inch lengths were carefully arranged to construct a tepee over the kindling. A dry match gave our tiny flame its start.

Attaining Intimacy

I was impatient as he studiously added twigs just marginally bigger, delicately leaning them into the little flaming pyramid. I tried offering twigs for the fire, but mine were always too big, until I finally accepted his way, the fire's way of growing at its own pace, and the fundamental caution of not crushing out the precious flame with something big, heavy, and wet, even if it's only a quarter-inch in diameter. What is small and flickering grows better that way.

Soon enough we were warm beside our fire, joyously adding large branches of wet wood, hearing them sizzle, seeing them smolder and smoke, then burst into flame, and we had more wood drying nearby. We changed into our dry clothes, which had been kept onboard in waterproof bags. Then we heated water and drank hot tea. Warmth and well-being had returned to the world, and I had ample time to muse about the benefits of gentle, careful encouragement.

People have a fondness and fascination for fires in the wild, even if that wild is just the backyard. Once the flame is tamed and channeled in a stove or oven, it is less interesting, but get a charcoal grill going, and people can become focused and passionate about transforming the raw into the cooked, and generating the heat it takes to do it.

While grilling may appeal to most everyone, it is often a guys' sort of thing, and guys do not ask other guys or anyone else how to do something, just like guys do not ask for directions when they do not know where they are or how to get to where they are going. If you ask for directions, somebody might think you're lost. If you ask for instructions, someone might think you are not thoroughly competent. Guys just know how to do things, and if they don't, well, guys just pretend they do.

The grilling itself is immediately, intensely gratifying, especially when conducted outdoors at one's leisure (as opposed to in a restaurant kitchen). People invariably volunteer to participate in the grilling. Yet this is not a job to be given up lightly, as though it were something frivolous that anyone could accomplish, so hesitate a little before you agree to pass on the long-handled tongs. Assess the sincerity of the supplicant. To watch food being cooked over fire is a high honor: to tend, prod, poke, peek, to smell the aromas, to observe the colors shift toward brown. Life is happening. We're at the heart of things where the harvest is transformed into food.

Wonderful conversations spring up around the fire as the food sits, cooking. Stories, tales, gossip, movie reviews, secrets shared about devotions and betrayals. A deep, warm, intimate place: food cooking over an open flame in the company of friends.

Grilling Instructions and Recipes

I discovered charcoal-grilled vegetables while working at Greens Restaurant in San Francisco in the early eighties. Deborah Madison, the founding chef at Greens, had been inspired to grill vegetables while cooking for Nancy Wilson Ross at her summer home in the Adirondacks, and she specified that the kitchen at Greens should include a charcoal grill, even though we would not be serving any meat. What a simple, yet brilliant concept: Apply traditional cooking methods to vegetarian cuisine.

To get their charcoal going, restaurant grills often employ a gas hose, which is quite effective. At home I have employed a charcoal chimney to excellent effect. To fire up this cylindrical metal device, newspaper is inserted in the bottom and the charcoal goes on top. Use the one sheet of newsprint the instructions suggest, as the "two is better" approach means the paper won't have enough air to burn. I have also started a fire using paper and kindling on the lower grate and put the charcoal on the upper grate. Once this starter charcoal is going well I dump it onto the bottom grate and add more charcoal. Portable gas grills can also be used with quite excellent results.

Long-handled tongs will make this job a lot more doable, and you may also want to keep a good hot-pad glove handy.

Grilled Eggplant Salad with Roasted Red Peppers
Simple Mixed Vegetable Grill on Skewers
Grilled Asparagus
Grilled Fresh Figs
Endive Salad with Grilled Figs and Fire-Roasted Walnuts

Grilled Eggplant Salad with Roasted Red Peppers

This is a variation on a recipe in The Greens Cookbook. *People enjoy it so much, you can have up to one eggplant for every two people – that's how good it is. Although basil and arugula are the herb seasonings that are called for, I have also made this salad using fresh tarragon or cilantro. If arugula is not available, go ahead and use a whole bunch of basil.*

Serves 4 to 6 people

2 globe eggplant
Olive oil
2 red bell peppers
½ bunch fresh basil
½ bunch fresh arugula
6 cloves garlic, coarsely minced
2 shallots, finely diced
Balsamic vinegar
Salt
Black pepper

Slice the eggplant crosswise into pieces ¼- to ⅜-inch thick. Brush each side with olive oil. Grill them over charcoal (Are you still using that wonderful mesquite charcoal and deforesting northern Mexico? Now you can get Eco Char, which is made from walnut shells) until they are well-browned on both sides and bend easily when tested with tongs. The eggplant should not be *al dente*, so if in doubt take a slice off the grill, cut it open, and try it. When done, set the eggplant aside.

Cut the red peppers in half, remove the seeds and pith. Cut the halves in half, toss with olive oil, and charcoal-grill, turning occasionally, until the deep red color (and the texture) softens, and there are perhaps spots of black. Set them aside.

Rinse and spin dry the basil and the arugula. Remove the basil leaves from the stems and cut the arugula into 1-inch lengths.

When the eggplant and peppers are cool enough to handle, cut the eggplant rounds into quarters and the peppers crosswise into ½-inch strips. (If you've gotten large sections of black on the pepper skins or if the skins have blistered, you are welcome to remove the skins before slicing.)

Combine the garlic and shallots, and then mix into the vegetables. Toss with the basil and arugula. Season to taste with balsamic vinegar, salt, and freshly ground black pepper.

Simple Mixed Vegetable Grill on Skewers

The possiblities for vegetables on skewer are limitless, but this one is simple, with good colors and minimal preparation.

Serves 4 to 6 people

12 medium-large mushrooms (fresh shiitake mushrooms are excellent for this)
1 green bell pepper
18 cherry tomatoes
1 cup olive oil
½ cup balsamic vinegar (or rice wine vinegar)

Attaining Intimacy

2 tablespoons soy sauce

6 cloves garlic, chopped

1 teaspoon sugar

¼ teaspoon salt

Black pepper

6 (8-inch) skewers

Wipe off the mushrooms with a damp cloth or paper towel. Cut the pepper in half lengthwise and remove the stem, seeds, and pith. Then cut the pepper in half crosswise, and these pieces in thirds. Rinse off the cherry tomatoes and remove the stems.

Combine the oil, vinegar, soy sauce, garlic, sugar, salt, and pepper for the marinade and let the mushrooms and peppers sit in this for an hour, give or take. Drain off and reserve the marinade, and put the vegetables on 6 skewers – the green and red are particularly good next to each other.

Grill over charcoal for 3 to 5 minutes, turning over occasionally, until the corners of the peppers are browned and the tomatoes are blistered. Serve as is, or drizzle some of the marinade over them.

Grilled Asparagus

Do not overlook asparagus or the fresh figs that follow when pondering what to grill – amazing.

Several asparagus stalks per person

Same marinade as above or plain olive oil

Salt

Coat the asparagus with marinade or olive oil. Grill several minutes, rolling over now and again, until spotted with brown and somewhat limp. Remove, salt, eat.

Grilled Fresh Figs

A couple of figs per person, such as fresh Black Mission, especially, or Kadota
or Turkish

Grill the figs whole, until they are limp and succulent. When placed in a bowl
or on a platter they will exude a marvelous elixir. Don't miss it. Don't miss the
figs. Your guests may not understand how extremely good these figs can be.
Clue them in.

Or don't; there is a great salad that can be made with the leftover figs. It follows.

Endive Salad
with Grilled Figs and Fire-Roasted Walnuts

*I make this salad with leftover grilled figs, which have been cooked until they are
limp, juicy, succulent, and sweet. Don't light the grill just for this, because you
could bake the figs for 12 to 15 minutes at 375 degrees to sweeten and soften them.*

Serves 4 people

12 fresh, ripe black Mission figs, grilled or baked

½ cup walnuts

1 tablespoon white sugar

⅛ teaspoon salt

2 tablespoons lime juice

1 tablespoon honey

1 or 2 Belgian endives (about 6 ounces)

Cut the stems off of the figs and slice them in half or quarters lengthwise.

Roast the walnuts for 6 to 7 minutes at 375 degrees until they're toasty and
aromatic. Then put them in a skillet on the stove and sprinkle on the white

sugar and salt. Cook over moderate heat, stirring, until the sugar melts and coats the walnuts. Remove and set aside.

Whisk together the lime and honey.

To arrange the salad, cut off the base of the endive and separate into individual leaves. Place a fig on each leaf and walnuts on the fig. Drizzle the lime-honey over the top.

A second method is to slice the endive crosswise into ½-inch pieces and toss with the figs and the dressing. Garnish with the walnuts.

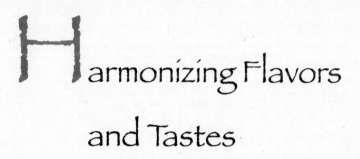

Harmonizing Flavors and Tastes

An Adventure in Dining Out

When I went to some cooking classes given by the "great chefs" of Szechwan, one thing that stayed with me was a description of the two most important points in Chinese cooking, although I couldn't help thinking there was something missing in the translation. The first principle was to visualize the dish: These are the ingredients; these are the seasonings; this is what makes the dish what it is. The second was referred to as "timing," but I think that what was meant was "execution" or "enactment."

That description seems rather good to me. Part of cooking is conception, visualization, dreaming, imagination, creativity. The other part is work: attention, care, skill, thoroughness, actually handling the stuff which is food. To wit, a story.

Many years ago I went out to dinner with three friends. We drove many miles to eat out at a cooking-school restaurant, and we arrived hungry and in eager anticipation of a fine meal. The restaurant felt warm and inviting, and after being greeted hospitably we were escorted to an upstairs table. In buoyant spirits we ordered a bottle of champagne and perused the menu.

Our education in the two principles of Chinese cuisine commenced with the first dish to arrive at our table, a cream of mussels soup. The first bite was delicious. By the third spoonful the excessive amount of salt was painfully distasteful. Four of us, hungry as we were, couldn't finish a single serving of this soup.

Floating in the bowl was a large crouton topped with grated cheese. "Perhaps," we conjectured, "the crouton is meant to be an antidote to the excess salt." Wrong. The crouton with its grated Italian cheese was saltier than the soup.

Someone's "timing" had been off, or perhaps their attention or taste buds. We weren't sure.

We turned our attention to the Salmon Bouillon, which made its appearance with a sprig of dead parsley floating on its surface. We knew the parsley was dead because it was bobbing belly up and floundered this way and that. Once we removed this unappetizing specimen the soup turned out to be delicious. Conceptual flaw we decided. Paper-thin slices of lemon or strands of zest might have been an alternative, or perhaps narrow shreds of seaweed. We let it go, and awaited the main course.

All the entrees were well-conceived and so smoothly executed that I cannot recall them in the slightest. Since all of them were indeed suitable and satisfying, we decided to order for the table the one entree that none of us had dared to try: Lobster Poached in Sauterne and Cinnamon. Frankly, this just hadn't sounded appetizing to any of us, but what did we know? Perhaps it was a classic combination of which our unschooled intelligence was unaware. Perhaps a mysterious alchemy took over the ingredients producing an inconceivable delicacy.

The lobster arrived, and each of us took a bite. We had been right in our initial reaction: Bad Idea. One of my companions described it as lobster in Maypo syrup. Once he said that, the idea stuck: Lobster in Syrup did nothing for the lobster.

By now we had had another bottle of wine or two in addition to the champagne, so we proceeded in high spirits determined not to be dismayed by anything strange still to come. Following the entrees, the most beautiful, carefully arranged salads arrived. Smaller leaves had been artfully placed face up inside larger ones, while calendula petals and pistachios dotted the surface along with fresh herbs. We felt grateful and delighted that someone would do this for us, until the first bite produced a sensation of grittiness between the teeth and around the tongue as did every succeeding bite of that stunning salad. Someone had been negligent in washing the lettuce. A simple thing, making all the difference in the world.

Harmonizing Flavors and Tastes

We were curious to know what had happened, so we asked the waiter, who returned to inform us that it was the end of the semester, and all the cooks were out on the veranda drinking champagne and congratulating one another.

Next came dessert. When we decided we would like a dessert wine, we could not resist trying an Australian one the waiter described as salty. We wondered what he could mean, until we actually tasted it: salty, a dessert wine that was salty. That was decidedly strange, but we hoped the desserts would bring atonement.

My genoise, thin slices of cake layered with buttercream, looked simply marvelous. Reverently I picked up my fork and gently moved to cut off the tip of the wedge of cake. Nothing happened, so I pushed down harder and harder until the cake sprang up at acute angles on either side of the fork vigorously forming a "V" shape. The cake was impenetrable by fork.

We asked for steak knives, and over espressos and cappuccinos we admitted our ignorance. We didn't know how a dessert wine could be salty, or why a genoise became cardboard. We didn't know if there had been a conceptual mistake or a flaw in executing the idea. We did know that we had had a thoroughly entertaining and enlightening evening.

When the check came, our waiter was not off in the slightest.

Recipes

My food generally emphasizes careful attention, rather than dramatic conception, although I do think things through and imagine various flavor combinations and whether or not they come together for me.

People can get lazy with one or the other. The work of preparing food suffers when we get unduly fascinated with the creative aspect, and we can remind ourselves that it is the food itself that is the Marvelous Creation, beyond anything we might conceive. Cooks, after all, don't create tomatoes. Yet thorough and careful work alone cannot make up for neglecting to consider what goes with what, what the key points are, which differences make a difference. So I try to take care of both aspects, while being grateful for the ingredients with which I am working.

This menu is a bit more elaborate than some, and your effort could bring it to fruition, making it what it is — vividly present, aromatic, and alive in front of you.

Radiatore with Mushrooms, Almonds, and Goat Cheese
Avocado Platter with Nectarines and Roasted Red Pepper
Sautéed Zucchini

Radiatore
with Mushrooms, Almonds, and Goat Cheese

Radiatore are little "radiators," which are marvelous for soaking up flavor. If un-available, pick another kind of pasta with lots of surfaces, like fusilli (spirals). For mushrooms I like to use a combination of cultivated white mushrooms and fresh shiitake or portobello, or if they show up as a gift from someone, I include chan-terelles or morels.

Serves 4 people

4 ounces almonds (about ¾ cup)

1 tablespoon olive oil

Medium-small red onion (about 6 ounces), sliced

¾ pound mushrooms, sliced

3 cloves garlic, minced

8 ounces radiatore or other pasta

½ cup sherry

Salt

Freshly ground black pepper

1 teaspoon balsamic vinegar

3 to 4 ounces goat cheese (chèvre), crumbled

Harmonizing Flavors and Tastes

¼ cup flat-leaf parsley, minced

3 to 4 ounces Parmesan or Asiago cheese, freshly-grated

Start heating 2 quarts of water in which to cook the pasta.

Roast the almonds on top of the stove in a dry skillet or in a 350-degree oven for 7 to 8 minutes until browned and crunchy. Slice or chop or use the Cuisinart on pulse.

Heat the olive oil in a large skillet and sauté the onion for a couple of minutes, then add the mushrooms and garlic and continue cooking. This is probably a good time to start the pasta cooking.

Once the mushrooms are cooked, add the sherry and let it cook down. Season the mixture with salt, pepper and balsamic vinegar.

When the pasta is cooked, drain it in a colander, and add it to the skillet along with the almonds and goat cheese. Mix well and adjust the seasoning. Garnish with the parsley, and serve the grated cheese on the side.

Avocado Platter
with Nectarines and Roasted Red Pepper

You must have noticed by now that I am a big fan of avocados and roasted red peppers. Here they are combined with nectarines: pale green with orange and brick red; oily smoothness and refreshing juiciness accented with smoky fleshiness.

Serves 4 people

1 Roasted Red Pepper (see recipe page 81)

1 to 2 tablespoons balsamic vinegar

Salt

Black pepper

2 avocados

2 ripe nectarines

⅓ cup olive oil

3 tablespoons orange juice

1 tablespoon lemon juice

2 teaspoons honey or white sugar

1 shallot, diced finely

12 to 15 leaves basil or mint *or* 2 tablespoons fresh tarragon *or* parsley, minced

Prepare the Roasted Red Pepper, cut it into strips, and season it with the balsamic vinegar, salt, and black pepper.

Prepare the avocados in decorative slices (see page 64). Fan them out on a large serving platter or individual plates.

Slice the nectarines. I like to try out different ways of cutting. One interesting way is to leave the nectarine whole, cut round slices off one side, and then the opposite side, then finish by pushing out the pit and cutting what's left. Arrange the nectarine slices with the avocado.

I prefer to pile the red pepper slices in 2 or 3 places, rather than distribute them over the avocado and nectarine.

Whisk together the olive oil with the orange juice, lemon juice, honey, and shallots, and spoon it over the avocado and nectarine slices. Sprinkle on some salt and pepper where you think it will do the most good (more on the avocados?). Slice the fresh basil or mint leaves into thin strips, and scatter them over the top.

VARIATIONS

If orange and lemon are not handy, try rice vinegar or champagne wine vinegar.

Sun-dried tomatoes, perhaps 8 to 10 of them, could be used in place of the roasted red pepper. Plump them, if necessary (see page 18), slice them, and put some dressing on them as well.

Another variation uses mango in place of the nectarine, and either roasted red peppers or sun-dried tomatoes. To cut up the mango, start by cutting off the skin. Then, holding the mango on its side, make slices until you reach the

Harmonizing Flavors and Tastes

pit. Then cut off slices from the opposite side. If you are willing to use odd-shaped pieces, cut off what's left of the mango flesh and slice it.

Sautéed Zucchini

When I prepare sautéed zucchini, I would rather not have zucchini mush. I prefer it tender, but somewhat al dente. *That to me is the secret of the dish.*

Serves 4 to 6 people

2 pounds zucchini
2 tablespoons olive oil
Salt
1 teaspoon dried basil *or* oregano *or* thyme *or* 1 tablespoon fresh herb, minced:
    parsley, thyme, or basil

Trim off the ends of the zucchini, and cut them in Chinese rolling-cut (pencil-sharpening) style: Start by making a diagonal cut on the zucchini – cut straight down with the knife nearly parallel to the zucchini. Then roll the zucchini slightly away from the knife. Make the next diagonal cut about a 45-degree angle to the first cut. Each cut piece will have some dark green from the outside of the zucchini and two cut surfaces. These cut pieces sauté better than rounds cut directly crosswise, because they do not stick together the way rounds do.

Heat a large skillet until it is quite hot – drops of water sizzle when scattered in the pan. Add the olive oil and a sprinkling of salt, then the zucchini, and stir-fry over high heat. Taste after 2 to 3 minutes, and add more salt if necessary. Then add the dried herbs, if using them. Cover the pan, reduce the heat, and let the zucchini steam until it is tender. Garnish with the fresh herbs, if using them.

Tasting the Fruit of the Vine

When I worked at Greens Restaurant I was invited to join a wine-tasting panel. Once a month or so, we would get together, sit down at lunch, and taste six to nine different wines with foods from our Greens menu. We had quite an enjoyable time, and I recall those afternoons with fondness and pleasure.

Our tasting panel included Dick Graff, who had started Chalone Winery; Alice Waters of Chez Panisse; Shirley Sarvis, who conducted food and wine tastings; Bob Finnigan, who wrote a food and wine review; Barbara Barnhardt, who worked in Dick's office; and several of us on the Greens staff, usually a couple of managers and one or two of the cooks: Deborah Madison, the founding chef, or Jim Phalan, her assistant. When I went to my first tasting I was in awe of my companions and nervous about whether or not I would be credited with having "good" taste. I needn't have worried.

We would taste each wine without food and then each wine with each dish. If you wanted to, you could spit out the wine you had tasted in a paper cup, but still being fairly young I just drank everything. After each taste I would make some notes, and with no formal wine-tasting vocabulary, I would just make it up: sharp, acidic, fruity, straw, varnish, spring flowers, heavy, light, robust, thin, asphalt, bell-like. I didn't have a twenty-point system or a hundred-point system. I just tasted and came up with the whatever words I could.

Besides making tasting notes we would rank the wines in our order of preference. Afterward we would total the numbers, and then unveil the bottles to see which was which and how we had ranked the wines. What surprised and delighted me was that there wasn't a "right" answer. The "experts" did not agree; I was entitled to my opinion. Taste was a matter of taste.

After we had removed the wines from their brown bags, we discussed them, and I would have the opportunity to hear what words others used to describe the wines. Gradually my vocabulary, and my ability to distinguish flavors, increased.

Harmonizing Flavors and Tastes

After a while I was asked to be the wine buyer and was given the responsibility of organizing the tastings. I soon found that most of the wines salespeople brought to the restaurant as samples did poorly in our tastings and that I had to track down good wines for myself. Once I did that, and the wines were of generally recognized high quality, then the results of our tastings were more variable than ever.

One tasting was particularly illuminating, since every one of the zinfandels tasted was ranked first or second by one taster and eighth or ninth by another. Each wine was well made, with good balance and no obvious flaws. Still one person would comment with distaste that a particular wine tasted like straw, while someone else would say how pleasant he found that same quality, so reminiscent of the dry California hillsides in summer with their scattered oak trees and late afternoon sun. One person's delight in the fruitiness of a wine was another's Kool-Aid.

I found that the words I used to describe the flavor of various wines could be divided into three main groups: earthy, herbaceous, and fruity. For me this categorization has become quite useful and evocative, corresponding to the three main parts of plants: roots; stems and leaves; flowers and fruits. And I use it now in reference to foods as well as to wines.

Recipes

If you would like a wine with the following menu of Mexican polenta and spinach, you might try a dry sauvignon blanc or a dry pinot noir vin gris or white zinfandel. These are cleansing and refreshing with the spiciness of the food. Alternatively, the fruitiness of a zinfandel could be quite complementary – that, or perhaps the grassiness of a merlot.

Polenta Mexicana
Simply Spinach

Polenta Mexicana

My friend Dan Welch in Santa Fe started the Spaghetti Western Cafe. I am uncertain whether or not it has a specific location. He is a big fan of genuine Italo-Ranchero cooking, which he has been pioneering. He knows about empowering himself to find his own way.

Polenta Mexicana is clearly in that vein – something Italian converted to Mexican. I encourage you to check it out. I pour the cooked cornmeal into a muffin tin to make decorative rounds for serving – one of the keys to polenta is to make an attractive presentation.

Serves 4 to 6 people

3½ cups water

⅛ teaspoon salt

1 cup coarse cornmeal (polenta)

1 recipe Tomato Chili Sauce (see recipe page 244) or other tomato
   sauce (pages 243 – 46)

½ red bell pepper

3 ounces (orange) Cheddar cheese

3 ounces Jack *or* Muenster cheese

2 ounces canned whole green chilies, cut in strips

½ cup Parmesan *or* Asiago cheese, grated (optional)

1 tablespoon fresh marjoram or oregano, minced, *or* several sprigs of cilantro or
   flat-leaf parsley

Heat the water to boiling, add the salt, then slowly whisk in the coarse cornmeal so that the water keeps boiling. Lower the heat and continue cooking and stirring with a whisk for several minutes until the mixture is quite thick. I let it go at that, rather than cooking for the more traditional 30 minutes, which is more important if you want to serve pillowy, soft polenta straight from the pot.

Get out a muffin tin with deep forms, and distribute the polenta among the muffin cups. (It is not necessary to grease the tin.) When the polenta has cooled, run a knife around the edges to pop the polenta "muffins" out of the tin.

Harmonizing Flavors and Tastes

While the polenta is cooling, prepare the Tomato Chili Sauce. Preheat the oven to 375 degrees.

Cut the red pepper into strips, blanch them in boiling, lightly salted water for a minute, and drain. Slice the two cheeses into strips.

To prepare for serving, ladle a cup of the sauce into a 9 x 13 x 2-inch baking pan and heat the rest of the sauce separately on the stove. Place the polenta "muffins" on top of the sauce, and drape the cheese, red pepper, and green chili strips on top of the "muffins." Bake covered in a 375-degree oven until hot, about 25 to 30 minutes.

To serve, ladle ⅓ cup of the sauce onto individual plates, and place 1 or 2 polenta "muffins" on top of the sauce. Garnish with the Parmesan cheese, if using, and fresh herbs.

Simply Spinach

After I had been cooking for a year, I tried to find out how to prepare spinach, simply spinach. I found recipes for spinach quiche and spinach pie, all very well and good, but what if I just want spinach? I was at a loss for what to do. One recipe said to fold the leaves in half and pull the big stems off the back of the leaves. We tried this, but with four cases of spinach to prepare we didn't get very far before someone suggested that the stems were not that tough. Sure enough, they are not too tough to eat, but the dish is more delicate if you remove them.

I find this plain spinach to be a good accompaniment to the polenta dish.

Serves 4 people

1 bunch of spinach (about 1 pound)
2 tablespoons sesame seeds, for garnish
Salt

A thorough cleaning of the spinach is vital. I start by cutting the clumps of spinach crosswise at the base of the larger leaves and sectioning these into

2-inch lengths. Then I take the stems and cut them crosswise about an inch above the roots. This allows me to sort out the smaller leaves interspersed among the stems. Wash the leaves by immersing them in large amounts of water. Usually at least two rinsings are necessary. Set the leaves aside in a colander to drain.

Toast the sesame seeds in a dry skillet over moderate heat, stirring as needed to keep them from burning, until they are crunchy and aromatic, perhaps 5 to 8 minutes.

Layer the spinach into a saucepan, sprinkling each layer moderately with salt, then cover, and cook over moderate heat. The water on and in the spinach is enough to keep the spinach from burning on the bottom of the pot, but if you are nervous, add a touch more. The spinach will wilt down in 2 to 5 minutes.

Check the seasoning and serve garnished with the toasted sesame seeds.

If you want to embellish Simply Spinach further, you might consider a touch of vinegar or lemon juice, black or red pepper, or curry spices.

Tomato Ecstacy

Almost everyone has the capacity to taste, to discriminate between various flavors, yet having the capacity doesn't mean that people exercise it. One reason it is underutilized is that people tend to be timid about using language to articulate the differences they have noticed. I find it fascinating how language helps develop taste. Often, when we cannot put a label on what we've noticed, it loses its significance. Conversely, when awareness has labels to attach to experience, suddenly details and nuances are relevant; they can be tagged.

One example of this is professional tea tasting. According to an article I read, almost anyone with the appropriate training can learn to be a tea taster. Participants at tea-tasting school are given twenty different teas and told,

"This is what we mean by 'bright.'" Even though the twenty teas are different, they have this one common characteristic, "bright," which the tasters are expected to identify. Another twenty teas are "bold" or "smoky," "chesty" or "full-bodied." In this way one can learn the requisite language and subsequently be able to pick out the "bright" or "bold" which gives Lipton's or Twining's its distinctive flavor.

Outside the context of a particular profession with its specific terminology, we are often at a loss. In response to the question "What is the flavor of a tomato?" we are likely to throw up our hands in exasperation – "Well, you know, it tastes like a tomato!" as if that explained it, and in our culture it often does. Growers market pale red objects that are shaped like tomatoes but have a mealy dry texture with the flavor of mildly tart water, and still they make a killing. The buying public doesn't find the distinctions important enough to give them voice.

Our culture teaches us that food is only food, a tomato is a tomato (no matter how bland and insipid it is), and we learn not to pay attention to what is most important: the essential vibrancy of tomato. When we fail to notice the essential juicy, lush, and meaty vibrancy of tomato, somewhere inside us our "heart" shrivels up, our succulent fecundity is unrecognized and uncalled for. We too are dry and mealy, and longing for something to break us open and make us feel alive and flowing.

In late summer I go to the Real Foods market on Stanyan Street in San Francisco and I buy tomatoes: Beefsteak tomatoes, Golden Jubilee, Lemon Boy, Striped Marvel, Zebras, Purple Cherokee, Green Grape cherry tomatoes. They often have a dozen kinds or more – red, yellow, orange, purple, and golden tomatoes – some sweeter, some more lemony, some meaty, fruity, herbaceous, or earthy.

I must tell you, there is a tomato-eating ecstasy (and it's completely legal). My mouth explodes with sunlight, water, blue skies, patches of cloud. Birds call and insects hum. Earth, be it red, black, brown, or yellow dirt, has been distilled into flesh and seed, skin and juice. My body responds and comes alive.

A smile breaks forth. I am home, a place wild and robust. If a tomato can be this fully a tomato, it must be OK for me to be fully me, with all my bugs and weeds and unexplored but fertile mud.

If a tomato is just a tomato, well then, you will never know this ecstasy, and somewhere inside will be unfulfilled appetites aching and yearning to be fed. Tomatoes with no discernible taste, pulpy pink water, will not satisfy this hunger for vigor and vitality. You can taste the difference.

All these tomatoes at Real Foods also have a "family" name, the place they come from, because that is also what or who they are: Knoll Farm, Full Belly Farm, Webb Ranch, Hungry Hollow, Terra Firma. When Kofi, a yoga teacher friend, came to dinner at our house, he said that when he grew up in Ghana they always knew where the fruits or vegetables were from, this hillside or that valley, and who grew them. It's unavoidably part of the vegetable, part of the fruit. He said that even after years of living in the West he still couldn't understand how we could eat anonymous produce so indiscriminately. Don't we have any sense about these differences?

What is not "measurable" tends to be overlooked. What is distinctly human, distinctly individual, unique, alive, different tends to be unacknowledged, unvoiced. I say, let the tomatoes sing, let them dance, let them do cartwheels in your mouth, let them awaken your heart, your soul, your spirit. Let them speak sermons, soliloquies, and sonnets.

Recipes

Here are some basic approaches to give voice to tomatoes.

Celebrating Tomatoes!
Garnished Tomato Platter
Tomato Salad with Avocado and Black Olives

Celebrating Tomatoes!

To celebrate tomatoes keep your eye out for tomatoes worth celebrating, whether from the supermarket, vegetable stand, or farmers' market, your garden or a friend's. The more flavorful the tomatoes are, the simpler I make the dressing, so that their unique gifts will not be overshadowed. The recipe that follows is a starting point, followed by some possible variations.

Serves 4 to 6 people

2 pounds tomatoes of your choice or what you can find
2 shallots, finely diced
⅓ cup rice vinegar
2 tablespoons white sugar
Salt
Black pepper

Rinse off the tomatoes and remove any stems. When I have more than one variety, I like to cut them and arrange them on a platter, keeping each kind separate, but you might prefer to mix them. Slice the tomatoes or cut them into wedges. Cut the cherry tomatoes or Sweet 100's in half, because cut surfaces release more flavor. Arrange on a platter.

Combine the shallots with the vinegar and sugar, and spoon over the tomatoes. Just before serving sprinkle on some salt and pepper. (Putting on the salt earlier will draw a good deal of water out of the tomatoes and make them soupy.)

Garnished Tomato Platter

Here are some ingredients that might enhance your tomatoes. Choose what you like from among them.

Serves 4 to 6 people

1 recipe Celebrating Tomatoes! (see recipe page 237)

2 to 4 tablespoons olive oil

Fresh garlic, just a *hint* (if you can't resist)

½ pound fresh mozzarella (or possibly feta), cut into slices or strips

⅓ cup Niçoise olives

¼ cup roasted pine nuts

1 tablespoon fresh thyme or marjoram, minced

1 dozen basil leaves, cut into strips

The olive oil or garlic can be combined with the dressing to spoon over the tomatoes. The fresh mozzarella (or feta) may be arranged with the tomatoes. The olives, pine nuts, or fresh herbs may sprinkled over the top of the tomatoes.

Tomato Salad with Avocado and Black Olives

In the summer I find I also enjoy the combination of avocados prepared with tomatoes. Here is a sample.

Serves 4 to 6 people

2 pounds ripe, juicy tomatoes of your choice (red, yellow, cherry)

2 avocados

1 bunch chives *or* ½ bunch green onions

⅓ cup black olives (Niçoise, Kalamata, or oil-cured) *or* 2 tablespoons capers

¼ cup olive oil

3 tablespoons red wine or sherry vinegar

1 teaspoon sugar

1 clove garlic, minced

Salt

Optional:

½ green chili (serrano or jalapeño), minced

1 to 2 tablespoons fresh marjoram or oregano, minced

Harmonizing Flavors and Tastes

Wash, de-stem, and cut the tomatoes into wedges or rounds and the cherry tomatoes in half. Cut the avocado lengthwise, remove the pit and the peel, and cut the flesh into fork-sized chunks. (Make them large enough so that they will not disappear into mush with some gentle stirring.) Cut the chives into ¼-inch sections, or cut the green onions into thin rounds, using as much of the green as you wish. What do you think? Are you going to pit the olives? Combine the tomato pieces with the avocado and olives.

Whisk together the olive oil, vinegar, sugar, garlic, and, if using it, the green chili. Toss lightly with the tomato mixture. Just before serving sprinkle with salt and garnish with the chives or green onion, and the fresh herbs, if using.

Root, Shoot, Flower, and Fruit

Flavors call out to me. I don't mean out of the blue, but I love to taste things. It's a way of knowing, a way of meeting, also a way of dissolving this too, too solid world. When I taste I go somewhere else, a world without crowds and stress, just you and me, and all the time in the world. I focus when I put something in my mouth. Something inside me melts. I am transported and completely still.

On one hand I simply enjoy, allowing my awareness to resonate with the various sensations. On the other hand, because I work with food, and talk about food and how to work with food, I also want to be able to articulate my experience. For this purpose I use not only the more classical categorizations of sweet, sour, bitter, salty, and pungent (or peppery), but also my own set — the "earthy" of root, the "herbaceous" of stem and leaf, the "fruity" or "vibrant" of flower and fruit.

Earth flavors can be deep, woody, smoky, or woodsy. Mushrooms remind me of mulch and meadow, of sojourns in the forest. Grains have a sweet earthiness, ranging in my experience from the intensely hearty and earthbound buckwheat to the more sunny corn and millet, and their colors correspond.

Bran and germ, like specks of dirt, flavor the more neutral quality of white flour or white rice.

Once I decided to taste plain lentils before all the other vegetables and seasonings were added for soup. Immediately I thought of dirt and fell in love with the earthy flavor, marvelling that a growing thing could make dirt so palatable. Yes, beans are in this category.

Potatoes, yams, sweet potatoes, carrots, beets – they are all root vegetables. Washed and scrubbed, and especially when baked, here is mellow ground, often sweet, occasionally bitter. I feel supported, rooted, nourished: deep red-purple earth, orange-brown ground.

The meat of land animals is largely in this category of earth: meaty, smoky, robust, dense, and chewy. Dairy products, especially cheeses, have earthy characteristics. In France some cheeses are described as "excremental."

As well as sweet earth there is also bitter earth: unsweetened cocoa, coffee, tea, nuts. Walnuts, almonds, sesame seeds, sunflower seeds – rich, oily, bitter earth.

Leaf, stem, stalk – lettuces, spinach, chard, kale – these flavors are herbaceous, grassy and green, along with green beans, green peppers, mustard greens, peas. As a group these are not as sweet as the grains and root vegetables. Here more is tart and bitter, pungent and acrid. Asparagus and celery epitomize stalks. Here also are broccoli, cauliflower, and the summer squashes. Tomatoes, although a fruit, seem "viney" and herbaceous rather than flowery or fruity. Eggplant, certainly, and cucumber, too.

These are flavors that move and leap in air and sunlight, flavors that engage the tongue. The effect in the body is also cleansing and invigorating. The body juices flow – motion that is called life-force blooming, or well-being. Earth by itself, meat and potatoes, will tend to become heavy, slow, at times too damp or too dry, while the flavors of leaf and stem help to regulate and harmonize the flow, draining the damp, moistening the dry, moving what is solid.

Onions and garlic represent a particular or distinctive branch of this category. Raw, and sometimes boiled or steamed, they can have a strong, stimulating pungency. When sautéed or baked they become quite earthy, mellow, grounding – a vital ingredient in so many dishes because they provide a bass note, a stabilizing element, a sweetness which moderates and enhances the vigorous flavor of "vegetable." Try, for instance, adding sautéed onion to plain tomato sauce (tasting before and after the addition), and notice how the singular note of "tomato" becomes a chord: rounder, fuller, more resonant.

Flower and fruit: here is a flash of color, the accent of lemon or orange. Apples and pears, peaches, and plums: the flavors in this category become jazzier. There is sweetness and also tartness and color. The flavors "dance": That layer of raspberry in the chocolate cake, the twist of lemon peel in the espresso.

All those berries in the summertime – the intensely colored stain-the-counter, dye-the-hands flash of strawberry, blueberry, blackberry. You know that sunlight has brought these to fruition, along with generous amounts of water and some earth. Melons are also in this category, although some which are not so sweet, especially cucumber, have more of the flavors of stem and leaf.

Particular seasonings are also commonly used to add this "high note" or vibrant quality to otherwise pedestrian dishes: vanilla in chocolate chip cookies (all those earthy flavors are suddenly a lot brighter), cinnamon in apple or pumpkin pie, a fresh basil pesto with pasta or pizza.

I distinguish the same trio of flavor qualities in herbs. Rosemary and sage clearly have woody, resinous characteristics which are most often too overwhelming for leafy greens and other "stalky" vegetables yet quite suitable for seasoning more "earthy" ingredients such as grains, beans, and potatoes or Thanksgiving stuffing. Parsley, marjoram, thyme, and oregano are in the midrange of herbs – herbaceous, pungent, bitter – they will "darken" a dish as well as "brighten" it. Parsley for example has a lemony side, but still won't enliven the way fresh lemon does.

Especially important in my cuisine are the fresh herbs that have flowery, bright, "fruity" flavors: basil, tarragon, mint, and cilantro. Adding any one of these makes prosaic ingredients or dishes start to "sing" in the mouth. The taste buds wake up and take note. For this purpose I also use lemon peel or juice, occasionally orange peel, vinegars, or any of an assortment of peppers: fresh green chilies, dried red chili powders, fresh ginger.

Spices fit into these flavor categories as well. For my taste onion seeds and fenugreek, two spices used in Indian cooking, are the earthiest. Cumin seeds, coriander seeds, and nutmeg are in my middle category, while allspice, cloves, cinnamon, fennel seeds, anise seeds, cardamom, and mace qualify for my vibrant, brightening classification. Often just a hint of one of these will liven up the flavors of a dish without being recognizable as "cinnamon" or "cloves" unless someone is particularly sensitive. All too often these spices are overutilized, and there is nothing but the flavor of "nutmeg" or "cinnamon."

Often when planning a menu or wondering what might enhance a particular dish, I turn to this flavor categorization. Could this lettuce salad use something earthy like roasted sesame seeds or grated Parmesan cheese or shall I brighten it with a vinaigrette utilizing anise seeds or cardamom? If the menu is not complete, what is missing – something substantial and chewy (and earthy) like a wild rice and brown rice pilaf? Something bright, juicy and refreshing? Green leaf or stalk? Within dishes, as well as among them, for me some balance or range is most pleasing.

Naturally science is finally getting around to studying these things. A friend of mine just told me about the results of a recent study, and guess what? Food that is attractive and pleasing is more nourishing than food that isn't. I've believed that all along. Science, though, isn't sure; it was after all a rather small sampling.

Recipes

For my cooking classes I have created a way for people to develop a sense of these flavor categories. We make four different tomato sauces, tasting the tomatoes at the start, and then proceeding to taste after each ingredient is added.

The sauces utilize dried red chilies of various sorts, but whether or not you can procure these, the interesting thing will be to get a sense of "earthy," "herbaceous" (stem/leaf), or "vibrant" (flower/fruit). To the extent you want to, you can create your own flavor-study class when you are making the tomato sauces.

I have, by the way, conducted canned tomato taste-offs at my classes over the years, and for a long time our champion was Bonnie Hubbard (the cheapest), beating out Hunt, Heinz, Contadina, Progresso, and S & W, among others. But Muir Glen organic canned tomatoes now top them all. Congratulations, a fine product. Be sure to get the plain whole tomatoes and not a flavored product; you can do that part better yourself!

Ancho Chili Sauce with Roasted Garlic
Tomato Chili Sauce with Roasted Sesame
Dark Tomato Sauce with Chili Negro and Cocoa
Herbed Tomato Sauce

Ancho Chili Sauce with Roasted Garlic

Baking gives the garlic a mellow quality in this sauce, while oregano makes the flavors more vibrant and alive. Notice also what happens when the small amount of vinegar is added. I think the flavors are brightened.

Makes 4 cups

1 head garlic (2½ to 3 ounces)
1-pound, 12-ounce can whole peeled tomatoes
About 3 tablespoons ancho chili powder (see page 19)
About ½ teaspoon dried oregano
About ½ teaspoon balsamic (or other) vinegar
Salt, if needed
Pinch of sugar, possibly

Start the garlic baking in a 350-degree oven. There are two ways to do this. I usually separate the cloves and bake them (unpeeled) on an oiled baking sheet, but another great way is to cut off the base of the garlic head, brush the head with olive oil, and bake.

When the garlic is quite soft and spreadable, about 25 to 35 minutes, remove from the oven and let it cool. Now remove the garlic paste from the skins. I open each clove and remove the pasty insides. If you have baked the garlic with the top cut off, you should be able to squeeze the roasted garlic out the end of each clove. (It's sometimes useful to keep a bowl of water nearby to rinse your hands, as this can get quite sticky.)

Blend the tomatoes, or coarsely chop them, and prepare the chili powder, if you have not already done so.

Try the tomatoes, then add chili powder to taste, starting with perhaps 2½ tablespoons. Puree the roast garlic with some of the sauce, then add it to the full pot. Taste and notice what you will: I call it a "deepening." The sauce has a denser, sweeter, heavier quality.

Next, notice what the oregano does. I aim to bring up the flavors without making the sauce taste like "oregano." Next, add the vinegar, and assess once again.

Do you want more of anything? Any salt needed? The pinch of sugar will "smooth" things out.

Tomato Chili Sauce with Roasted Sesame

This sauce is enriched with roasted, ground sesame seeds. The fresh cilantro brings up the spiciness and awakens the taste buds. In Mexican cooking a sauce with ground nuts or seeds is known as pipiano. *This one is quite simple.*

Makes 4 cups

¾ cup brown sesame seeds
1-pound, 12-ounce can whole peeled tomatoes

3 to 5 tablespoons New Mexico red chili powder (see page 19)

¼ to ⅓ cup cilantro, minced or cut into thin strips

½ teaspoon balsamic vinegar

Salt, if needed

You may want to treat this, too, as a cooking class and a chance to educate your palate by adding one ingredient at a time and tasting after each addition.

Roast the sesame seeds in a 350-degree oven for about 8 minutes until they are aromatic and crunchy. Let cool a few minutes, then grind in a coffee (or spice) mill.

Blend the tomatoes or chop them coarsely, and start them heating on the stove. Prepare the chili powder. Mix the chili into the tomatoes starting with 3 tablespoons and adding more to taste. Add the ground sesame seeds, which will give the sauce a toasty nuttiness. Now stir in the cilantro, taste, then add the vinegar. Want more of anything? Do you think it needs salt?

Dark Tomato Sauce with Chili Negro and Cocoa

This is my bare-bones version of a mole sauce, using unsweetened cocoa. More than just "earthy," this sauce is distinctly "dirty." The touch of cinnamon makes the flavor "sparkle" just a bit. See what you think.

Makes 4 cups

1-pound, 12-ounce can whole peeled tomatoes

2 tablespoons powdered chili negro (see page 19)

1 tablespoon unsweetened cocoa

1 to 2 pinches of cinnamon

½ teaspoon balsamic vinegar

Bit of salt, perhaps

2 to 3 pinches sugar, if you wish

Blend or coarsely chop the tomatoes with the juice in which they are packed and start it heating. Prepare the powdered chili negro, and add it to the tomatoes to taste, starting with ½ tablespoon. Add the unsweetened cocoa, again to taste, and then the cinnamon. Then add the vinegar.

See how you like the sauce. Consider whether or not you might like a bit of salt or sugar.

Herbed Tomato Sauce

Of these four tomato sauces this one is the closest to a traditional European tomato sauce. Once again, if you wish to educate your palate, you can add the ingredients more or less one at a time, tasting after each addition. It is not quite the usual way to make the sauce, but it will work.

The sautéed onion and garlic "deepen" or "round out" the flavor of tomato. The acidity of the wine has an effect similar to that of the vinegar in the recipes above: The flavors become "brighter." The herbs contribute flavor complexity and help to liven up the sauce, especially the fresh mint or basil. See what you think.

Makes 4 cups

1-pound, 12-ounce can whole peeled tomatoes
1 tablespoon olive oil
1 yellow onion, diced
2 to 3 cloves garlic, minced
½ cup red wine (or dry white)
2 tablespoons parsley, minced
¼ teaspoon dried thyme *or* 1 teaspoon fresh thyme, minced
Salt
Black pepper
2 tablespoons minced fresh mint, fresh basil, *or* fresh tarragon

Blend or chop the canned tomatoes, juice and all, and start the mixture heating. (If you are chopping the tomatoes by hand, strain off and reserve the juice, chop the tomatoes, then recombine.) Once the tomatoes are warmed up some, see how they taste.

Heat a skillet, then add the oil and let it heat. Sauté the onion for 2 to 3 minutes, then add the garlic and sauté for another 2 to 3 minutes. Add to the tomatoes and taste what happens.

Pour the red wine into the still-greasy, oniony skillet, and let it cook down by half while scraping the bottom of the pan to incorporate the caramelized sugars. Then add the wine to the sauce, and taste what that does.

Add the parsley, then the thyme. How is it?

The canned tomatoes are often well salted, so you may not need any more, but see what you think. Next, season with the black pepper, and finally the fresh mint, basil, or tarragon.

By now you should have some sense of the contribution each ingredient is making, so you can consider what if any adjustments you would like to make. (You might also consider a spoonful of fragrant olive oil, a dollop of vinegar, or a pinch of sugar.)

If fresh mint, basil, or tarragon is unavailable, try adding ¼ teaspoon *herbes de Provence.*

Visualizing a Meal

Visualizing a dish or a meal means imagining how the various ingredients fit together and whether or not there is a pleasing or appetizing congruence of elements. I find this similar to the zen teaching: "Let your mind go out and abide in things; let things return and abide in the mind." This is not about using fewer ingredients in order to simplify, or adding more elements in order to elaborate. I understand it as "connecting" or "relating intimately," the work of a lifetime, never finished, yet already complete.

To help me visualize I use both my own division of flavors into "earthy," "stemmy," and "fruity," and the more traditional listing of five tastes – sweet, sour, salty, bitter, and pungent or peppery. Asian cuisines often add a sixth taste to this set, known as "plain." Incorporating plain taste can make the organizing principle for a meal fundamentally different.

In my cooking I have tended to move away from the model of Western cuisine, in which the meal is usually structured around a main event or entree, with preliminary events leading up to it. Perhaps by definition the warm-up acts are mediocre, so that by comparison the main attraction is much more powerful and appealing. "At last," you say, "something I can sink my teeth into. This was worth waiting for." Instead of creating a main event I enjoy weaving more of a fabric of flavors, tastes, and textures.

In some ways this is related to the Asian paradigm of organizing the meal around the plain taste, frequently in the form of rice. A meal is then like going on various flavorful adventures and then returning home to rice, bread, or a simple grain dish. Home is plain and dependable. Here you can skip the surprises, please. Home is purposefully bland. From here you can go out and experience something salty, something sweet and sour, something bitter, something hot and spicy, and then return to what is stable, plain, enduring. And the plain provides a counterpoint that highlights the various flavors.

The adventure might also include crunchy, chewy, soft, dry, juicy; red, green, white, or orange. The menu is not a matter of building up to a main attraction, but of providing a variety of engaging dishes. Of course you want to eat again in two hours. Your digestive system has not been overwhelmed into sluggishness. Thank goodness.

When you think about it you can see that the Eastern approach is potentially much more interesting. Instead of a big piece of meat, with every bite tasting the same, you could have two curries, raita, and three chutneys, plus roasted coconut and cashews. Depending on how you combine things, that's more than twenty-seven flavor combinations, and you are just getting started. Every bite of your meal could be different from every other.

Harmonizing Flavors and Tastes

Here is a simple dinner I prepared that featured potatoes, tomatoes, and asparagus. We can visualize and examine the various elements and their balance. See what you think.

The red potatoes are baked in a puddle of red wine with salt and pepper. When the potatoes are quite tender, the pan is uncovered, a modest amount of cream is added, and then baking continues (uncovered) until the cream thickens slightly. For visualizing let's look at the various ingredients and the seasonings: What are the colors, the flavors, the textures? Do they all fit together or not? What is the key to the dish, the pivotal point that makes it what it is?

The potatoes here are smooth and creamy, and this is heightened by the velvety quality of the cream. It's also notably pink, with its pink skins and pink sauce, and it has the sweet succulence of baked rather than the sweet dry of fried, which calls for ketchup. The sweetness of the potatoes is enhanced by the sweetness of the cream and set off by the mild tartness of the wine and the pungency of the freshly ground black pepper. Potatoes baked this way have a marvelous earthiness – the flavor of purified, refined dirt. Can you taste it yet? Do you have it in your mouth?

To complete this "picture" – but what is a picture painted with flavors? a taste-fest – we will need tart, juicy, refreshing and crunchy, chewy. We'll want something to cut the butterfat, something to chew, some stem and fruit, air, water, sun.

A tomato salad offers tart and palate-cleansing juiciness. Here is the fruit of sunlight, ripe and fleshy, which contrasts with the more stodgy, earth-bound potato. The tomatoes are graced with fresh herbs, including parsley, scallion, and thyme, as well as with a touch of balsamic vinegar and a hint of olive oil. Thinly cut slices of provolone cheese set off the redness, and its earthiness brings a hint of the first dish to the second.

Here the tartness of the tomatoes is accented by the vinegar, softened by the sweetness of the cheese: sun-earth, tart-sweet. And the herbs are pungent. Where the potatoes are a solid chord, the tomatoes sing melody.

The asparagus is stalk, herbaceous and energetic, the green at the interface of heaven and earth. It is robust – the reaching of the potato, the support of the tomato. While sturdy and supple, the asparagus is also lithe and tender, its flavor bittersweet. Cutting the stalks into long diagonals and then cooking them in olive oil and a bit of butter over high heat brings out more of their sweetness and also preserves some of their crunchiness. Garnishing with some slivered roasted almonds provides added nuttiness, and the short sets off the long. The other dishes being essentially soft, this chewiness becomes an important addition.

A handsome plate: spray of asparagus dotted with almond, mound of potato, spread of tomatoes and cheese. Beautiful colors, pleasing aromas – though completely ordinary, the meal also feels absolutely unique. It's *just this*: not too exciting, not too dull, something with which to connect and resonate.

All this is nothing like mishmash, the practice of adding more ingredients to disguise and cover up a basic lack of attentiveness to the object and its characteristics, an unwillingness to ascertain and harmonize the various qualities. An uncalm mind will be unable to appreciate the virtue inherent in the object. The cook who practices mishmash will never know the Tao of the kitchen.

And so I encourage you to study these foods, to see, smell, taste, and touch these foods, enjoying their infinite characteristics. Enough talk! Let's eat.

Recipes

What follows are the recipes for the meal described above. I think it exemplifies what I mean by saying that my cuisine is not especially elaborate. I hope you'll enjoy it too.

Potatoes Baked with Wine and Cream
Tomato Salad with Provolone and Fresh Herbs
Asparagus Sautéed with Roasted Almonds

Potatoes Baked with Wine and Cream

This is a recipe to take your time with. It's so very easy, yet so delectable, and the important factor is time. You get to start the potatoes baking, and because the rest of the menu has a short preparation time, you can go out shopping, watch TV, read a book, or arrange for a leisurely session of lovemaking. Meanwhile the potatoes will take care of themselves.

In the end the potatoes will be creamy, mellow, smooth, melt-in-your-mouth. This is really the kind of dish for which you don't have to measure anything, but I went to the trouble of measuring everything just to give you an idea.

For 4 to 6 people

2 pounds red potatoes (about 10 medium-sized)
Head of garlic (perhaps 25 cloves)
1½ to 2 cups red wine
Salt
Black pepper
¾ cup cream

Preheat oven to 375 degrees.

Wash the potatoes and cut them into chunks. (Usually it will do to cut the potato in half and each half into 4 pieces.) Peel the garlic and cut the largest cloves in half. Place potatoes and garlic cloves in a casserole dish or baking pan. Pour in the wine so it comes about halfway up the potatoes. Sprinkle with salt and pepper. Cover. Bake.

Bake for a good hour and a half or longer. If it ends up being 2 hours, that's probably fine too. If you are around and think of it, you can stir the pot now and again, enjoy the developing bouquet, and return it to the oven.

After the minimal 1½ hour baking, stir the potatoes and add the cream. If you are trying to be modest about butterfat, you could try adding a little less. Uncover, and continue baking another 20 to 30 minutes while you prepare the rest of the menu.

That's about it. Oh, you might check the seasoning and add a grinding of black pepper on top. Most important though, allow your awareness to savor the potatoes and be soothed and delighted.

Tomato Salad
with Provolone and Fresh Herbs

Serves 4 people moderately

2 pounds fresh, ripe tomatoes

4 ounces provolone cheese

3 to 4 green onions, thinly sliced (both whites and greens), about ½ cup

¼ cup flat-leaf parsley, minced

2 tablespoons fresh thyme, minced

2 tablespoons olive oil

2 to 3 tablespoons balsamic vinegar

Salt

Black pepper, freshly ground

Cut the tomatoes in half vertically. Then cut out the stems and cores, and cut the tomatoes into wedges. Place in a bowl. Cut the provolone cheese into thin strips and mix with the tomatoes. Toss with the green onions, the herbs, the olive oil, and some of the vinegar.

Since the salt will draw water out of the tomatoes, wait to add it until right before serving. At that time season with salt, black pepper, and perhaps additional vinegar.

Asparagus Sautéed with Roasted Almonds

Serves 4 to 6 people

¼ cup almonds
2 pounds asparagus
1 tablespoon olive oil
Several pinches of salt
½ teaspoon grated lemon peel

Roast the almonds in a 350-degree oven for 8 minutes until toasty, or in a dry skillet over moderate heat. Let cool, and then slice them. (Already slivered almonds will not provide the same flavor; somehow they always taste stale to me.)

Snap off the tough ends of the asparagus by hand. Then cut it into 3-inch-long diagonal strips.

Heat a large skillet and add the olive oil and a sprinkling of salt. Sauté the asparagus for 2 to 3 minutes. Taste, then cover and cook over low heat until tender enough for your taste. The asparagus should still be bright green.

Toss with the lemon peel, check the seasoning, and serve, garnished with the almonds.

Please Enjoy Your Food

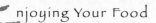

Enjoying Your Food

At my Saturday meditation retreats, when we break for lunch, I often tell people, "Please enjoy your food." All morning I have been offering various instructions in sitting and walking meditation, and by lunchtime we have also had an hour of yoga with further directives, so I may leave it at that. I don't want eating to be another chore, or yet another place to worry about whether or not you are doing it "right." We do enough of that already, so I want to invite people to simply "please enjoy your food."

Occasionally I might say a bit more, although I don't want people trying too hard to have fun. I explain that enjoying your food is very important, because by enjoying something we connect to the world, to one another, to our inner being. When you enjoy your food you will be happy and well nourished by what you eat.

Sometimes I also explain to people that by enjoying their food, they will naturally find themselves practicing meditation. They will be paying attention to what they are eating, noticing flavors and textures and nuances of taste, because to enjoy something you need to experience it.

Also they will have to stay present, because if they get carried away by greed, they will miss what they are eating in the present, while thinking about the future possibilities. Entering into full enjoyment, they will be relaxing and opening their hearts to the food, not worrying about good and bad or right and wrong. The question of "how well am I doing this practice" will not come up.

Mostly I think it's better to say as little as possible. Then enjoying your food may be the best meditation you do all day. It takes care of itself without your having to try too hard.

Following the path of pleasure is deep and profound, and richly rewarding. Sometimes people complain that it doesn't work that way and that one needs discipline and restraint ("or I'd be a blimp!"). That's nonsense the way it's usually understood, implying that one's inherent being lacks wisdom or any sense of beauty and consequently needs to be kept in line and tamed.

Most of the problems that arise in the pursuit of pleasure are due to lack of devotion – not being fully enough committed to pleasure. Which bite of chocolate cake is no longer pleasurable? Which swallow of wine brings you down instead of up? Sure, restraint is needed, but it comes after pleasure or along with pleasure, not before and in place of pleasure.

When pleasure or enjoyment is forbidden, then we look for stupor or unconsciousness, which is the closest we can get to relief from the inane drive to discipline and restraint, and the overriding admonition not to have any fun.

Please enjoy your food.

Recipes

When I cook, I pay more attention to making my food delicious and enjoyable than I do to using or not using particular ingredients, although that practice may be quite helpful for some people.

I firmly believe that giving one's attention or awareness to the experience of eating will naturally change what one eats, and also that the effort to cook meals oneself, rather than eating prepared foods, tends to develop awareness.

I've never been to cooking school, but over the years I keep picking up things as I go along. One evening after we were divorced, I had dinner at my ex-wife's house. Her future husband, a physicist from France, made dinner: crêpes for every course. A variety of fillings was available: ham, cheese, tomatoes, vegetables, seasonings, and for dessert, ice cream and fresh berries. As the crêpes came off the stove, we could assemble them with our chosen filling: "Let's see . . . do I want the berries in the crêpe and the ice cream on top, or the ice cream in the crêpe and the berries on top?"

Alain explained to me his "1-2-3" crêpe recipe: 1 cup of flour, 2 cups of milk, 3 eggs, a pinch of salt, and, voilà, crêpe batter. Since I am not usually cooking for that many people I often make my batter with ⅓ cup flour, ⅔ cup milk, and ⅔ (one) egg. Sometimes I end up adding a touch more flour to make the batter slightly thicker.

I still like and make the asparagus crêpes in The Greens Cookbook, but I continue to try out various fillings. Here the crêpes are filled with mushrooms, another one of my favorites, and paired with carrots.

Mushroom Crêpes with Mushroom Sauce
Carrots with Roasted Sesame Seeds

Mushroom Crêpes with Mushroom Sauce

Make the crêpes first, then the filling, then the sauce. When all the ingredients are ready, then you can do the final assembly and serve-up. Though not as simple as rolling up some handy ingredients, this dish makes a lovely presentation.

Nowadays a wide range of mushrooms seems to be readily available. Here in the markets in California are fresh shiitake mushrooms, portobellos, brown field mushrooms, and oyster mushrooms in addition to the generic white mushrooms. I enjoy using some of these more flavorful and chewy mushrooms in combination with the more common less expensive variety. With the shiitake and portobello mushroom remove the stems before slicing.

Makes 12 or more 6 to 8 inch crêpes

Crêpes (see recipe page 261)

1 pound mushrooms, sliced

2 tablespoons butter

6 cloves garlic, minced

A few pinches of dried thyme

Salt

1 medium yellow onion, diced

1 tablespoon olive oil

2 tablespoons parsley, minced

A few pinches of dried basil

Mushroom Sauce (see recipe page 262)

Greens of 6 green onions, thinly sliced diagonally

Black pepper

Prepare the crêpes, following the recipe on the next page.

To make the filling, sauté the mushrooms in the butter for 3 to 4 minutes – you may need to do this in two batches – until the mushrooms have started to brown, then add the garlic, thyme, and a couple pinches of salt. Cook another few minutes, until the juices have come out. Set aside.

Sauté the onion in the olive oil until softened, and then add the parsley and basil. Combine with the mushrooms and drain, reserving the liquid (which I call elixir) for the sauce.

Make the Mushroom Sauce, following the recipe on page 262.

For assembly and serve-up, place a crêpe on the counter with the smooth brown side down, and spread some filling across the middle. Roll up each crêpe around the filling and place in an oiled pan. Brush the tops of the crêpes with milk or water, and heat in a 400-degree oven for 12 to 15 minutes.

To serve, ladle about ½ cup of sauce onto individual plates, and place a crêpe or two on top. Garnish with the sliced green onions and freshly ground pepper.

CRÊPES

Makes 12 or more crêpes

⅔ cup unbleached white flour

1⅓ cups milk

2 eggs

Touch of salt

Touch of butter

Put the flour, milk, eggs, and salt in a bowl and whisk until smooth, or close to it. Use a smooth-surfaced, nonstick frying pan, 6 to 8 inches in diameter. Usually a medium to medium-high flame works best, but you can adjust it as you go. Put the pan over the flame and let it heat up.

I usually butter the pan the first time around, and then after that, it is not necessary if the pan is still in good shape. Have the pan hot enough so that a few drops of water sizzle when tossed into it.

To start the crêpe, pick up the pan with one hand and pour in about 3 tablespoons of batter with the other hand. Immediately swirl the batter around in the pan so that it completely coats the bottom.

Two to 3 tablespoons of batter will make a 6-inch crêpe, 3 to 4 tablespoons will make an 8-inch crêpe. Extra batter may be poured off back into the main batch. Holes in the crêpes may be filled with spots of batter or left for decor.

Let the crêpe cook for 45 seconds to 1 minute on the first side. Wait for it to show browning around the edges. This is a great time to use a heat-resistant rubber spatula: I get the edge of the crêpe up with the spatula, and then pull it up the rest of the way and turn it over by hand.

The second side finishes in about 15 seconds, since it doesn't need to brown. Finished crêpes may be piled up on a plate.

Be patient with the crêpe making; you may be eating a few imperfect ones before they start coming out well. If using butter in the pan, the crêpes will not brown evenly.

6 green onions, whites and pale green parts

2 stalks celery, finely diced

2 tablespoons butter

1 tablespoon olive oil

Salt

2½ tablespoons white flour

2 to 2½ cups milk

⅓ pound fresh shiitake (or cultivated) mushrooms, sliced

1 tablespoon olive oil

Black pepper

Trim the green onions, removing the rootlets and any wilted or limp stalks. Cut thinly, reserving the greens for garnish. Sauté the whites and pale greens of the onion along with the celery in the butter and olive oil. When softened add a sprinkle of salt and the flour to make a roux. Cook 5 to 6 minutes over moderate low heat, stirring occasionally.

Meanwhile add milk to the reserved mushroom elixir (from the recipe above) to make a total of 2½ cups of liquid. (If you are not making the crêpe recipe on page 261, use 2½ cups of milk.) Heat in a saucepan. When the flour is cooked, remove the roux from the fire, let the bubbling stop, then pour in the heated milk and whisk to combine. Return to moderate heat and cook until thickened, whisking occasionally.

Sauté the shiitake mushrooms in the olive oil until browned. Add to the sauce. If necessary, thin with more milk.

Season with salt and pepper.

Carrots with Roasted Sesame Seeds

I find that carrots can be a delicious accompaniment, when cut up and stir-fried. Here they have a light sweet and sour seasoning with some toasted sesame seeds.

Serves 4 people generously

2 tablespoons sesame seed
1 pound carrots
1 to 2 teaspoon olive oil
Juice of 1 lemon
2 tablespoons honey
Salt or soy sauce

Roast the sesame seeds in a dry skillet until they are fragrant and crunchy. Cut the carrots into julienne strips.

Heat a skillet, add the oil, and sauté the carrots for a couple of minutes. Add the lemon and honey, and cook, covered, over moderately low heat until tender. Season with salt or soy sauce.

Sprinkle the sesame seeds over the carrots and serve.

Becoming Intimate with Your Ingredients

Sometimes I wonder just how strange I am. Or are other people just as strange in their own way? Then again perhaps it's not really so strange to be having conversations with fruits and vegetables in the grocery store. How else can a person decide what foods to bring home? Doesn't it make sense to get to know them first? Putting food into our mouths is clearly one of the most intimate things we do. That stuff is going to become flesh and bone, thought and dream. I wouldn't want to get intimate with just anything.

Planning what to eat can be trying. Open a cookbook; close it. Think of something to eat; worry if it's good enough. Do the same old thing, tried and true. No, too boring. Try something new. No, too stressful. What turns out to be most helpful for me is to open the refrigerator or to look in the garden. Then I head for the store and see what talks to me, and what I want to have a relationship with.

Picking out what food to bring home is a mysterious business, about as easily explained as how one finds partners and friends. Sometimes we love food immediately without ever asking why – the way I took to my daughter, whom I hadn't even met until she was born. Now that's pretty amazing – to love someone so completely, regardless of how she looks, sight unseen, without even getting to know her. But mostly I'm a bit more cautious. I want to know what I'm getting myself into (or what's getting into me).

The type of conversation the food strikes up says a great deal about the nature of the prospective relationship. The supermarkets are full of packages that stand around looking pert, handsome, or sexy and whisper, "Buy me, buy me, buy me." If you agree, quiet approval and reassurance are in store: "You won't be sorry." "My, don't you look classy being seen with me?"

It's another story though, if you ask why you would want to buy them. The beautiful packages say, "I'm quick, I'm easy . . . I'm quick, I'm easy."

"Quick? Easy?"

"Sure, you won't have to relate to me at all. You won't have to think, feel, decide, or sense anything. Just put me in the oven or the microwave, and I'll be there for you." No problem. Use me and abuse me.

If you hesitate, things can turn ugly: "What's wrong with your taste, bozo, don't you know something good when you see it?" "What, you can't afford me? Are you sure you really belong here?"

With other foods, especially in the produce department, the pitch can be animated: "Hey dude, how you doing? Listen up, fella, you want to hear a great idea for salad? I'm dying to get together with that endive across the aisle. Help a fella out, would you? Man, we would dance and sing for you."

"Yeah, well, I was looking for a quiet evening."

"No problem, tone things down with the dressing."

"Okay, you're on."

Perhaps the yams notice how tired you are: "Been a long day has it? Well hanging out here in the supermarket getting squeezed isn't so hot either. How about if we stew together?"

The conversations are give and take instead of clever putdowns. These are foods that wouldn't mind if you touched them, felt them, smelt them – foods that will engage you and bring out the best in you. These are foods that acknowledge and respect you as a whole person, and not just as someone who has money to burn and won't be any trouble.

So when you're in the store, listen up and don't be afraid to have your say for fear of spoiling a tantalizing relationship. Do it with your mouth closed though; otherwise people might think you're strange.

Recipes

Here are a couple of recipes inspired by listening to the ingredients. These two fall or winter dishes utilize dried fruits for enhancing the flavor of the main vegetable, a combination I find appealing.

Yams Baked with Dried Apricots and Orange
Warm Red Cabbage Salad with Dried Fruit and Feta

Yams Baked with Dried Apricots and Orange

Serves 4 people

4 ounces dried apricots
1½ pounds yams (perhaps 2 large or 3 to 4 smaller ones)

Grated peel of 1 orange
Juice of 2 oranges
Salt

Preheat oven to 375 degrees.

Cover the apricots with water and cook them in a covered pot over moderately low heat until they are tender, perhaps 10 minutes.

Cut the yams crosswise into 1-inch sections. (Large yams may be cut in half lengthwise first.) Place in a baking dish or pan with the apricots and their cooking liquid, the orange peel and orange juice. Sprinkle on some salt. Cover and bake at 375 degrees for about 45 to 50 minutes, until the yams are tender. If you check them after 30 minutes and the yams are still in a puddle, you might want to leave the cover off the last 15 to 20 minutes to let them dry out.

Warm Red Cabbage Salad with Dried Fruit and Feta

This recipe takes a California cuisine classic and turns it into something less. Still, I enjoy poking around for what to use, besides the cabbage, in the way of fruit, nuts, and cheese.

Serves 4 to 6 people

½ cup sunflower seeds
1 teaspoon white sugar
Salt
2 tablespoons olive oil
1 red onion, diced
3 to 4 cloves garlic, minced
1 pound red cabbage, cut into thin shreds
1 teaspoon fresh rosemary, minced

4 ounces dried fruit (apricot, peach, or pear), stewed and sliced

2 tablespoons balsamic vinegar

4 ounces feta cheese (sheep's milk)

½ to ¾ cup Parmesan cheese, about 2 ounces, freshly grated

1 tablespoon fresh thyme, minced *or* 2 tablespoons parsley, minced

Roast the sunflower seeds in a dry skillet for 5 to 8 minutes until lightly browned. Sprinkle on the sugar and a pinch or two of salt. Stir briefly to dissolve the sugar, and then remove from the heat.

Heat the olive oil in a large skillet and sauté the onion for several minutes. Stir in the garlic, cook briefly, and then add the cabbage. Stir and cook a couple of minutes, then add the rosemary, dried fruit, and vinegar. Cover the pan, reduce the heat, and cook for 2 to 3 minutes until the cabbage is as tender as you like.

Stir in the feta cheese, and check the seasoning. Serve garnished with the Parmesan and the thyme or parsley.

Eating Just One Potato Chip

Years ago at a meditation retreat we had an eating meditation. Raisins were passed out. We were encouraged to help ourselves to a small handful, "but don't eat them yet!" I sighed. I am not thrilled with this kind of exercise. I prefer to have these experiences on my own, instead of having them spoon-fed to me.

We were instructed to look at the raisins, to observe their appearance, to note their color and texture, "but don't eat them yet!" I supposed it could be worse, like "ready now, one, two, three, open your heart to the raisins." Next we were invited to smell the raisins, and, finally, after a suitable interval allowing time for the aromas to register, we were permitted to put the raisins in our mouths, "but don't chew them yet!"

By now I was feeling annoyed and increasingly aware of an urge to smash something. "Leave me alone," I complained (loudly to myself). "Let me eat, for goodness' sake." To have your act of eating abruptly arrested is upsetting and disturbing. Get something tasty in your mouth, and your teeth want to close on it. But WAIT! We were then instructed to simply feel the raisins in our mouth, their texture, their presence. We were obliged to note saliva flowing and the impulse to chew.

At last we were permitted to culminate the act of eating. The raisins could be chewed. More juices flowed. The sweet and the sticky were liberated from their packets, "but don't swallow yet!"

"Be aware of your swallowing. See if you can make your swallowing conscious." Some people, I guess, just have a knack for knowing how to take all the fun out of things. This noting and observing, attending and awakening certainly doesn't leave much opportunity for joyful abandon, but I'll always remember those raisins.

Indeed I thought of them when I taught a workshop on Zen and psychoanalysis with Andre Patsalides, a Lacanian psychoanalyst. We called the event "Eating Orders and Disorders." Andre explained that in cultures where eating rituals were widespread, people experienced few eating disorders. Conversely, we see that ours is a culture with few eating rituals and numerous disorders. Many families, perhaps twenty-five to thirty percent, almost never eat together, according to many reports. The refrigerator, freezer, and cupboard are full of each family member's favorites, which can be microwaved when each one wishes, maybe between TV shows.

It's the American dream, the American way: freedom, disconnection, food as product, food as fuel, never having to interact. The basic rule, of course, is to not pay very close attention to the stuff – food, sit-com, people, or game show – coming in and then to be just a bit baffled as to why you feel so undernourished in the midst of all this plentitude.

I wanted to lead our workshop in an eating meditation, but hey, I thought, let's get real. Let's skip the raisins and meditate on eating just one potato chip.

Then I thought we could go on to oranges, my concession to wholesome, and conclude with Hydrox cookies. I picked Hydrox because I had heard they were the "kosher Oreos" (no pig fat, I guess).

Since I didn't want to parcel out the instructions as they had been given to me, I laid out the whole deal at the start: Pay attention. Allow your attention to come to the potato chip and be as fully conscious as you can of the whole process of eating just one potato chip. Just one! So you had better pay attention.

When I announced our potato-chip-eating meditation, I was greeted with various gripes, taunts, and complaints: "I can't eat just one." "That's ridiculous." "You're going to leave us hanging with unsatisfied desire. How could you?" Nonetheless I remained steadfast in my instructions and passed around a bowl of potato chips, urging each participant to take just one. When everyone was ready we commenced. "Instead of words," Rilke says in one of his sonnets, "discoveries flow out astonished to be free." And so it was.

First the room was loud with crunching, then quiet with savoring and swallowing. When all was fed and done, I invited comments. Many people had been startled by their experience: "I thought I would have trouble eating just one, but it really wasn't very tasty." "There's nothing to it." "There's an instant of salt and grease, and then some tasteless pulpy stuff in your mouth." "I can see why you might have trouble eating just one, because you take another and another to try to find some satisfaction where there is no real satisfaction to be found." "If I was busy watching TV I would probably think these were great, but when I actually experience what's in my mouth it's kind of distasteful."

That one potato chip even surprised me, the experienced meditator, with its tastelessness. Now I walk past the walls of chips in the supermarket rather easily without awakening insidious longings and the resultant thought that I really ought to "deny" myself. I don't feel deprived. There's nothing there worth having. And this is not just book knowledge. I *know* it.

The oranges were fabulous, exquisite, satisfying. The reports were: "Juicy . . . refreshing . . . sweet . . . succulent . . . rapturous." About half the participants refused to finish the Hydrox cookie. One bite and newly awakened mouths simply bid the hands to set aside what remained: "This we know to be something we do not need, desire, want, or wish for. Thanks anyway."

The ritual of eating attentively in silence put everything in order.

Recipes

First of all, I present a review of how to eat just one potato chip and then a recipe for tofu burritos, a dish we have fairly often in my house, unlike chips. Although tofu is not one of my favorite foods, I always find this burrito truly satisfying.

How to Eat Just One Potato Chip
Tofu Burritos

How to Eat Just One Potato Chip

Bets have been made. Challenges have been laid. You've been told you can't do it. You've never dared to try, but here's the secret. Taste. Taste what you put in your mouth. Experience it!

The potato chip is already manufactured and is always "ready for you" (waiting perhaps innumerable eons for this opportunity), so concentrate on preparing the other ingredients: To strengthen and focus the concentration, eliminate all the most obvious distractions: TV, radio, stereo, reading material (especially *People* magazine and the daily newspaper), talking, shopping, driving. Concentration is to be applied to the potato chip and only to the potato chip. No dip allowed. You are encouraged to be seated and not to have a drink in the other hand.

Attention is to be attuned to what is actually present moment after moment. "Attuned" because attention is often turned toward what is wished for or feared and frequently glosses over the actual experience. Refine or focus the attention by pointing out what is to be attended to: how the chip feels in the hand, how the chip looks in the hand, the smell of the chip, the intention to place said object in the mouth, how the chip feels in the mouth, how the chip tastes (moment after moment!), how the chewing sounds, and, carefully now, the sensations of swallowing.

Mindfulness is to be "whipped up" or aroused, as it tends to save itself for things more important than chips. Remind yourself that eating a potato chip with mindfulness is vitally important. To be mindful means that the experiences attended to actually make an impression.

One way to arouse mindfulness is to practice making notes about what you are going to tell your grandchildren about this particular potato chip: "beige… greasy between the fingers . . . exquisite curve . . . cute ruffles . . . urge (like a fire flaming to life) to place in mouth . . . feel with tongue . . . *powerful* crunch . . ." and so forth. But please, don't take my word for it. Find your own words.

Got your ingredients together? Seated? Undistracted? Focused? When you are ready you may pick up and eat (better yet, savor) that one potato chip. Get everything you can out of that chip, because it's the only chip in the entire universe.

Tofu Burritos

Burrito restaurants – taquerias – have become much more common in California than they used to be, so it has encouraged us to make more burritos at home too. It's also a way I use up leftovers (see page 110).

Sometimes my companion Patti cooks these burritos for me. Then the food is not only food; I feel valued and cared for.

As much as I might wish that chips could do that for me, they don't.

Makes 6 burritos

8 to 10 ounces tofu

½ tablespoon cumin seeds

1 tablespoon oil (or ghee)

1 small to medium yellow onion, diced

1 clove garlic, minced

1 teaspoon ground chili (see page 19)

2 stalks celery, diced

1 small green bell pepper, diced

1 carrot, diced

1 tablespoon parsley, minced

2 teaspoons dried oregano

1 avocado

6 regular-sized flour tortillas (rather than the gigantic ones)

Salsa

Crumble or mash the tofu and place in a strainer to drain off excess water.

Roast the cumin seed in a dry skillet over moderate heat until it is toasty and fragrant. Grind in a spice mill or extra electric coffee grinder.

Heat a large skillet, add the oil, and sauté the onion several minutes until it is translucent. Add the ground cumin, garlic, and ground chili, and cook another minute or so.

Then mix in the celery, bell pepper, and carrot. Cover, reduce the heat to low, and let the vegetables steam until they are tender. If the pan is unduly dry, add a spoonful or two of water to help with the steaming.

Once the vegetables are cooked, add the tofu and herbs, and cover to heat.

Cut open the avocado, remove the pit and skin, and cut it into slices.

To make the burritos, first heat the tortillas one at a time in a large skillet. Patti likes to fry the tortillas in a spot of ghee or oil, but I just heat them in a dry skillet.

Place the warmed burrito on a countertop and spread a portion of the filling across the middle, leaving the ends clear. Add 2 or 3 slices of avocado and a portion of salsa. If necessary pick up the near side and far side of the burrito to

Please Enjoy Your Food

help the filling form into a log shape across the middle. Then spread open on the counter, fold in both ends, fold the near side over the filling, and continue rolling the burrito over to enclose the filling. Place in a baking pan or another skillet for reheating, if necessary.

Prayer Helps

Throughout the day I offer many prayers as the occasion arises: "May you be happy, healthy, and free from suffering." "Just as I wish to be happy, may all beings be happy." "May you enjoy vitality and ease of well-being."

I am not asking for everything to be better, or for all your dreams to come true, but given that things are as they are and go as they go, I wish for your well-being and happiness in the face of all the changing circumstances. Things quite likely will not go ideally or according to plan, so I wish for the growth of buoyancy, flexibility, and resiliency. I wish for the nurturing of generosity and tolerance.

In the context of Buddhism I do not see prayer as necessarily directed toward a supreme being or higher power. Rather I see it as a clarification and expression of true heart's desire, or what my teacher Suzuki Roshi called innermost request.

What is it we really want? To know and act on true heart's desire or innermost request usually involves unearthing, sifting, and sorting. Speaking it can help to reveal and clarify it.

Each day I offer a prayer before meals. I like using an ecumenical expression: "We venerate all the great teachers and give thanks for this food, the work of many people and the offering of other forms of life." There are many possibilities: "May this food bring us health, happiness, and well-being." "Just as we have enough to eat today, may all beings have enough to eat." "May this food nourish us (me) body, mind, and spirit." It could be as simple as "Blessings on this food," or "We thank Thee for this food."

To have food on the table is truly a blessing, and one's life can change profoundly by acknowledging one's gratitude and appreciation. As Rilke suggests:

To praise is the whole thing! One who can praise
comes toward us like ore out of the silences of rock. . . .
Everything turns to vineyards, everything turns to grapes . . .

If you use your verse whenever you eat, even when snacking – it can be silent or spoken – it will help bring you into the present and will have a tremendous effect on how you receive your food and assimilate it. Acknowledging the blessedness of food is also acknowledging your own blessedness, your own capacity to nourish other beings as well as your self.

Nourishment comes from receiving food (or any experience), fully taking it in, assimilating what is useful, and letting go of what isn't. In Buddhism what comes into our lives is called Dharma, or teaching. In Christianity all that we receive can be viewed as a gift from God. Gratitude is called for: "We give thanks for this food, this 'teaching,' this 'gift.'"

Lately I have been reading Larry Dossey's *Healing Words: The Power of Prayer and the Practice of Medicine.* Dr. Dossey is a physician who began incorporating prayer into his practice of medicine after reviewing scientific studies that demonstrated its effectiveness. He found the evidence for the efficacy of prayer to be simply overwhelming, even though this is one of the best-kept secrets in medical science.

What he points out is that prayer works regardless of religious background or belief. Also, it turns out that the most powerful prayer is not one that aims for any particular result, but one that is more all-encompassing: "Thy will be done," or "May the best results occur."

Along with a blessing or grace before meals or snacks, other eating rituals can be beneficial. "Ritual" in this sense could include sitting down at a table to eat, rather than eating standing up, walking, or riding in an automobile. Another is to turn off the TV and radio and to eat in the company of family or friends, or to focus solely on eating rather than eating and reading, or eating and talking on the phone.

Each of us can determine which rituals are most helpful. In this sense ritual can be seen as ways to do things that help to heighten or deepen awareness. Noticing tastes, physical sensations, feelings, thoughts, and moods will inform or "enlighten" the food choices we make, and our capacity to be nourished by the food we are eating. Giving our attention to the experience of eating is powerful, whether we are eating wholesome foods or unwholesome foods, or are overeating.

Ritual, prayer, your innermost request – please find your own way to bring yourself to your meal, to sitting down at the table and taking the time to eat with gratitude, enjoyment, and gusto. May your endeavors bring you sustenance and strengthen your sense of connection with all life.

Recipes

Along with prayer, I cannot help feeling that desserts help. So with apologies to anyone on a fat-free diet, here are two fruit tarts. By their nature tarts tend to be more decorative than pies, and tart dough is much less tricky to make successfully than pie dough.

Strawberry Rhubarb Tart
Red Bartlett Pear Tart

Strawberry Rhubarb Tart

I find the utter redness of this tart to be engagingly bright and heart-lightening, as well as sweet and sour, bitter and juicy – all coming together to arouse gustatory pleasure. Where some people may see only butterfat, I see red without any conflicting thoughts.

Makes 1 (9-inch) tart, serving 4 to 6 people

2 pounds rhubarb

⅓ cup sugar, or more to taste

½ grapefruit, peeled, sliced, then coarsely chopped

Juice of ½ lemon

⅛ teaspoon allspice

4 teaspoons Campari

Tart Dough with Lemon Peel (see recipe page 277)

THE TOPPING

2 baskets fresh strawberries (about 1½ pounds)

¼ cup sugar

1 teaspoon anise seeds, freshly chopped

THE GLAZE

1 tablespoon maple syrup

2 teaspoons balsamic vinegar

Wash the rhubarb and cut off any bruised or green ends. Then cut it into 1-inch pieces, and put in a saucepot with the sugar, grapefruit, lemon, and allspice. Let it cook slowly over low heat, covered, until the rhubarb releases its juice. After about 20 to 30 minutes remove the lid, raise the heat some and let the rhubarb mixture cook, bubbling gently, until it has thickened into a paste. This may take 30 to 45 minutes. Be careful not to have the heat so high that the rhubarb burns. Once the rhubarb has thickened, remove from the heat and stir in the Campari.

While the rhubarb cooks, make the Tart Dough, following the recipe on page 277, and bake it empty about 20 to 25 minutes at 375 degrees, as directed.

Wash and then core the strawberries. (I use the blunt end of a vegetable peeler to do this.) The berries may be left whole, or perhaps the larger ones may be cut in half. Place the berries in a bowl with the ¼ cup sugar and the anise seeds. (I mince the anise seeds by hand with my knife – it's fairly easy to do.) Let the berries sit while the rhubarb cooks and the tart bakes.

After the tart has baked, allow it to cool in the pan for 15 to 20 minutes or more. Then spread the rhubarb mixture in the bottom of the tart shell. Drain the berries, reserving the liquid, and arrange them densely on top of the rhubarb.

Put the reserved liquid – perhaps a quarter cup? – in a small saucepot, add the maple syrup and balsamic vinegar, and cook briefly until it thickens. Drizzle this syrup over the strawberries. Remove the outside of the tart pan, and place the tart (on its metal bottom) on a plate for serving. Most excellent!

Tart Dough with Lemon Peel

1 teaspoon grated lemon peel

1 cup unbleached white flour

¼ cup whole-wheat flour

Pinch of salt

2 tablespoons white sugar

½ cup unsalted (sweet) butter, cut into 8 to 10 pieces or so

1 teaspoon vanilla extract

1 teaspoon water

Preheat the oven to 375 degrees.

Combine the grated lemon with the flours, salt, and sugar, and then cut in the sweet butter with 2 knives or a pastry cutter until a fine meal is formed. (A food processor may be used for this by pulsing.)

For tart dough it also works to use your hands. If right-handed, cup some flour and butter in both hands, then transfer it to your left hand with the palm up and fingers flat and outstretched. Move the heel of your right hand down and away against the mixture, pressing the butter flat between the fingers of your left hand and the heel of your right palm. Repeat as needed to make a mixture with a fine crumb.

Once the crumb mixture is formed, lightly toss with the vanilla and water. This is not intended to form a dough, but a "crumble." Using a 9-inch tart pan with a removable bottom, press the mixture evenly against the sides and bottom. I like to press it into the sides first, using both thumbs, being careful to make the bottom of the sides as narrow as the top of the sides. At first the dough should go up past the top of the sides, and then you can use your thumb to pinch off the dough so that it is level with the sides. Then use all the scraps and the remaining crumble to cover the bottom of the pan.

For the rhubarb tart, bake the shell empty of filling. As a precaution I sometimes poke a few holes with a fork in the bottom of the tart shell, but mostly my tart shells don't seem to puff up if I just bake them empty.

Bake the tart shell in a 375-degree oven for 25 minutes or so until it is aromatic and the sides of the tart have pulled away from the edge of the pan. The tart shell does not need to brown; it is actually a bit overdone if it does.

Red Bartlett Pear Tart

Buy the pears a few days in advance to give them time to ripen. I've made this tart in a variety of ways, but here is today's. A crumble topping underneath the pears soaks up their juices. You could always substitute regular Bartlett pears for the red Bartletts.

Makes 1 (9-inch) tart, serving 4 to 6 (or even 8) people

Tart Dough with Lemon Peel (see recipe page 277)

2 tablespoons sweet butter

2 tablespoons white sugar

¼ cup white flour

⅛ teaspoon cardamom

⅛ teaspoon anise seeds

2 to 3 red Bartlett pears

Please Enjoy Your Food

Make the Tart Dough and press it into a 9-inch tart pan as instructed but do not prebake.

Cut the butter and sugar into the flour along with the spices. Distribute over the tart dough.

Quarter the pears, core them, and then cut them into diagonal slices. Arrange them decoratively in the tart pan, fanning them out or placing them in concentric circles starting from the outside, or you figure it out. If you wish, sprinkle just a little sugar on top.

Bake in a 375- to 400-degree oven about 35 to 40 minutes, or until the sides of the tart are nicely browned and the pears are tender.

Gratitudes

Gratitude for Cooks

Knowing the labor and attention that goes into cooking, I feel an abiding gratitude for cooks, both those close and near and those far and wide; for all those who shop and wash, cut and grind, bake and sauté; and also for those who plant, hoe, and harvest. I extend this gratitude to cooks throughout time, an ancestral lineage coming down through the centuries: those who know which mushrooms are which and how to cure olives, those who fry, and those who stir.

Over the centuries such a tremendous effort has been made, endless hours of patient and impatient toil, the proverbial hunting and gathering, digging, skinning, shelling, all of which has brought us here today. Our bodies have fed on oysters and clams, pumpkins and corn, deer and quail, creatures and plants we might no longer recognize, but even more intimately our bodies have been fed by countless unacknowledged labors.

During hours and days in sun and rain, hands have become frozen, dried, cracked, lined, so that we may eat, so that we may drink. Looking at your hand, you may catch sight of this, but especially in old, well-used hands you see this, you know this: These hands have worked. They are knowing hands; they know how to cut grapes from the vine, how to rig the fishing line. The aged aunts' hands in Venice know how to flick gnocchi curls from the mass of dough, and the women laugh as our hands, which are so clever in their own way, fail to display the same easy dexterity.

Sometimes when we are quiet we feel it in our hands, feel it in our bodies, that effort and toil, that resourcefulness and resoluteness which has brought us here. Our bodies, our beings are full of it, replete with it. We aren't here by chance. Innumerable beings worked their hearts out to bring us here. The body is not just made of skin and bones and tissue, but is also made of this effort and caring of past generations.

Cooks everywhere embody this tradition and transmit this body to us. We arrive in this moment. Gratitude pours forth.

The Chinese Zen master Yueh-shan expressed his life like this:
"Awkward in a hundred ways, clumsy in a thousand, still I go on."

Thank you for your sincere efforts. I wish you good health, happiness, and well-being.

ndex

Note: My intention is to create a book with usable, accurate recipes. And I have attempted to correct some of the more obvious errors, which have crept in. Still, if you have difficulties with a recipe, or any questions or comments, you are welcome to write to me:

Edward Brown
c/o Riverhead Books
200 Madison Ave., 17th Floor
New York, NY 10016